TRIGONOMETRY MADE EASY

The Simple Workbook for Beginners

WiZO LEARNING

ISBN: 978-1-951806-31-6

Table of Contents

Introduction

At its most basic, the study of trigonometry is a study of triangles and their parts as well as the relationships between those parts. This includes looking at basic points, lines and angles and examining the relationships between a triangle's parts and how those relationships can be better applied to gain more information about the shape provided. In completing this course, students will be able to review concepts related to trigonometry starting with an algebraic review of coordinate plane concepts and ending with complex trigonometric laws. This includes looking at the distance formula and calculating distances between two points as well as using such laws as the Law of Sines and Law of Cosines to solve triangles for unknown quantities.

Along the way, students will gain knowledge of angles and arcs, the Pythagorean Theorem and right triangles and a familiarity with concepts related to the unit circle. Students examine the angle in standard position and then develop an understanding of radians and degrees and how to convert between the two measurements. Students look at parts of a circle and parts of a triangle and how circles and triangle measurements relate before looking closely at right angles. Students prove and use the Pythagorean Theorem to find a variety of unknown measurements and then gain an understanding on trigonometric functions. Students then apply the trigonometric functions to solve for unknown quantities in mathematical and real world situations. Finally, students gain an understanding of the unit circle. They use reference angles and graphs to examine trigonometric functions using unit circles and practice how to use graphs to describe trigonometric functions. Now that students have gained a full understanding of the functions, they apply them to trigonometric identities and equations and end with reviewing the various laws related to trigonometric functions.

This workbook introduces students to concepts, walks students through sample problems related to concepts and then has students apply the concepts as they solve problems on their own. There are additional review problems for each concept allowing students to gain true mastery in practicing the skills further. For success in completing this workbook, it is recommended for students to start by reading the content for each section. As they read, students should highlight main concepts or take notes. After reading the text, students should work through the example problems, following the steps provided in the workbook. After completing the sample problems, students should work through the practice problems in the section on their own. The review problems at the back of the book can be used for additional practice for each section or as a cumulative review once all sections have been mastered.

In completing their study of this workbook, it is hopeful students will gain an understanding and mastery of trigonometry. Once trigonometry is mastered, students can apply these concepts in a variety of professions or go on to study more advanced mathematics.

The Distance Between Two Points

We can use our mathematical knowledge to calculate the distance between two points. The best way to do so is to use our knowledge of the Pythagorean Theorem to find the length of the line formed by two points.

First, let's review the Pythagorean Theorem. The Pythagorean Theorem states that we can find the sides of a right angle triangle by using the following formula: $a^2 + b^2 = c^2$, where a and b are sides of the right triangle and c is the hypotenuese, or the side across from the right angle.

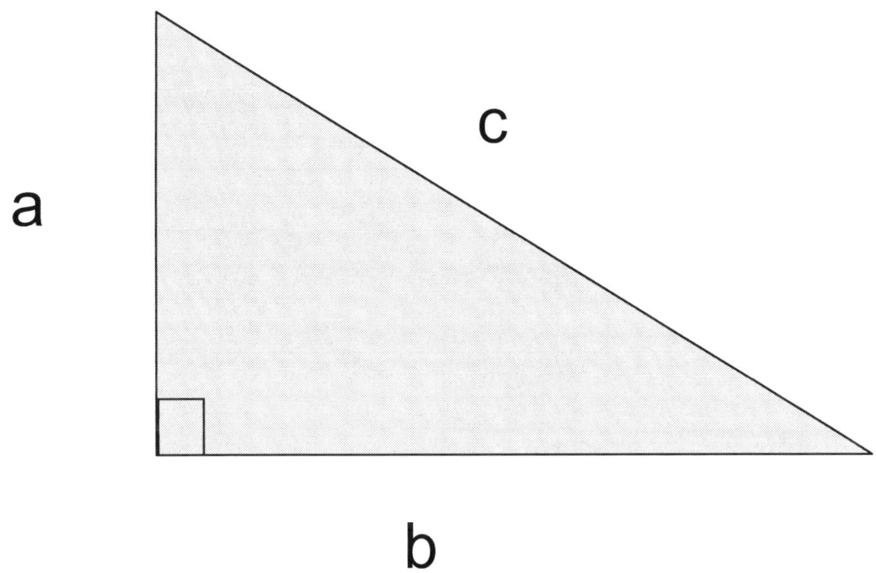

We can apply this theorem to finding the distance between any two points on the coordinate plane.

Let's look at an example on the following page.

Calculate the distance between (-4, 3) and (2, -1).
First, we should graph the points.

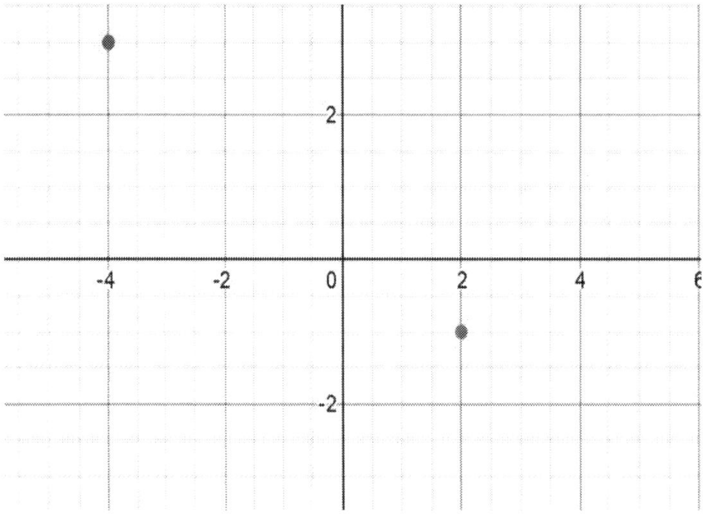

Now, we can turn the points into two parts of a right triangle.

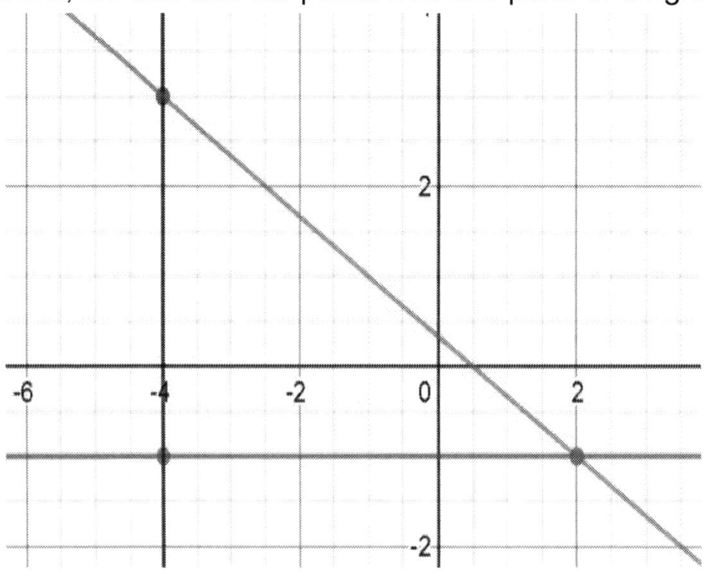

We want to find the length of the blue line above. To do so, we should use the Pythagorean Theorem.

Let's find the length of a and b first, as we are trying to find the length of c.

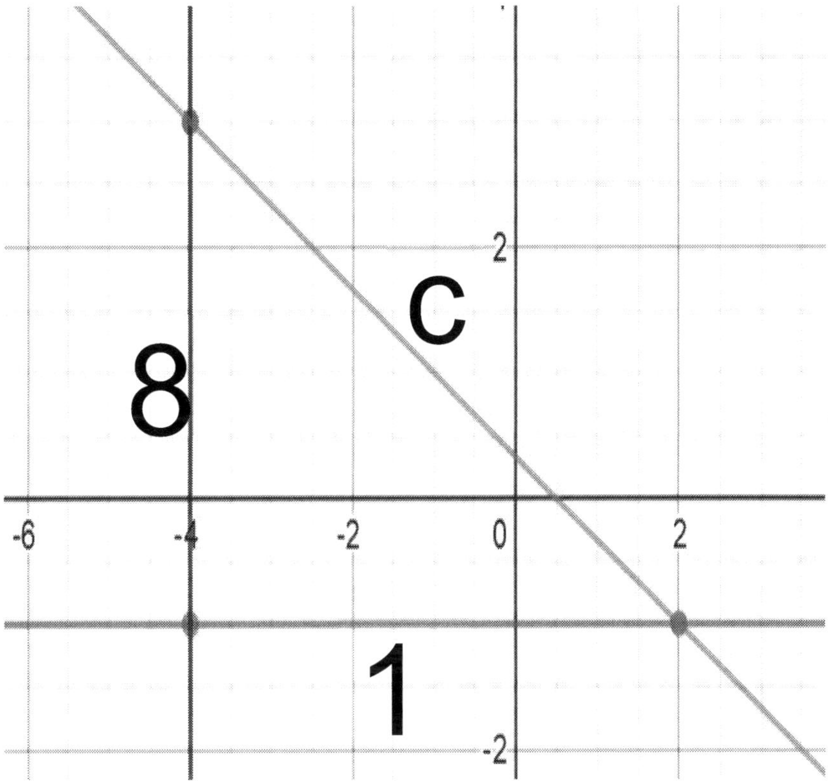

We use the formula to find c:

$$8^2 + 12^2 = c^2$$
$$64 + 144 = c^2$$
$$208 = c^2$$
$$\sqrt{208} = \sqrt{c^2}$$

14.422 is approximately c.

So (-4, 3) and (2, -1) are approximately 14.422 units apart.

Let's practice some more examples:
Example 1: What is the distance between (-6, -1) and (8, 14)?
Remember to graph your points first:

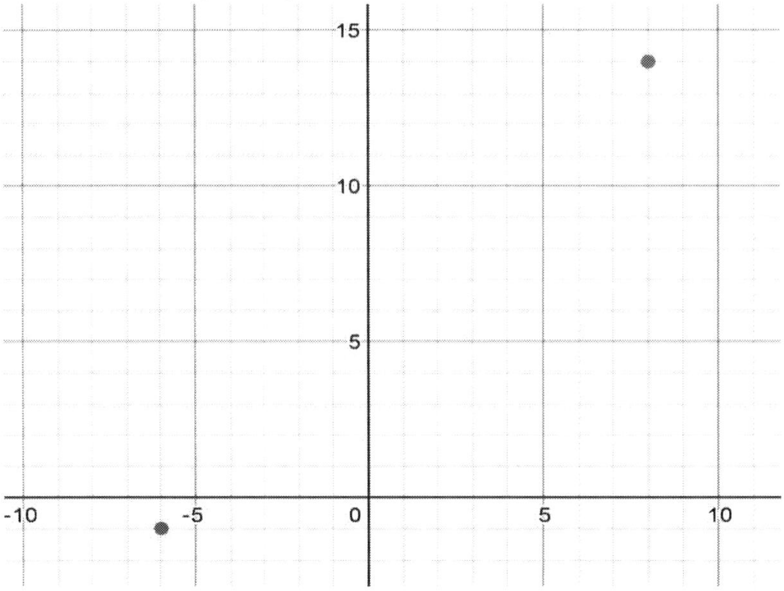

Then, turn your points into a right triangle.

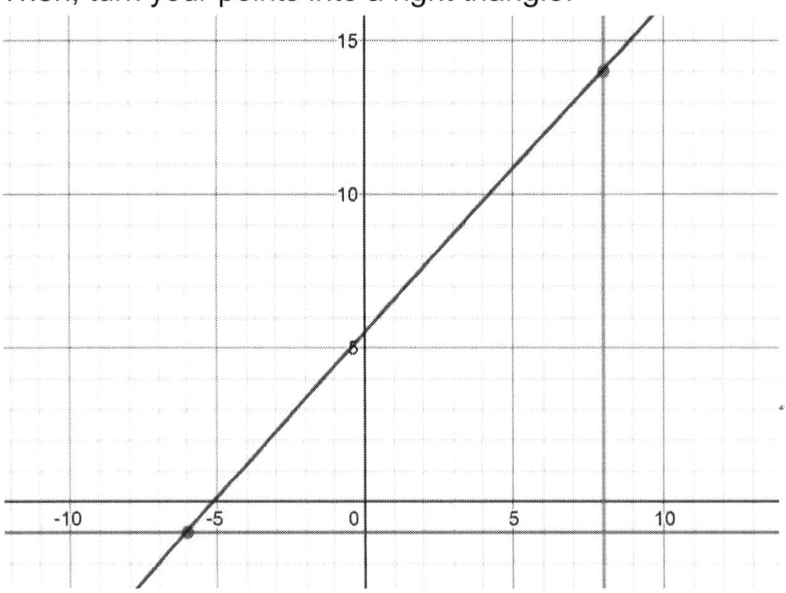

Now, find your values for a, b and c.

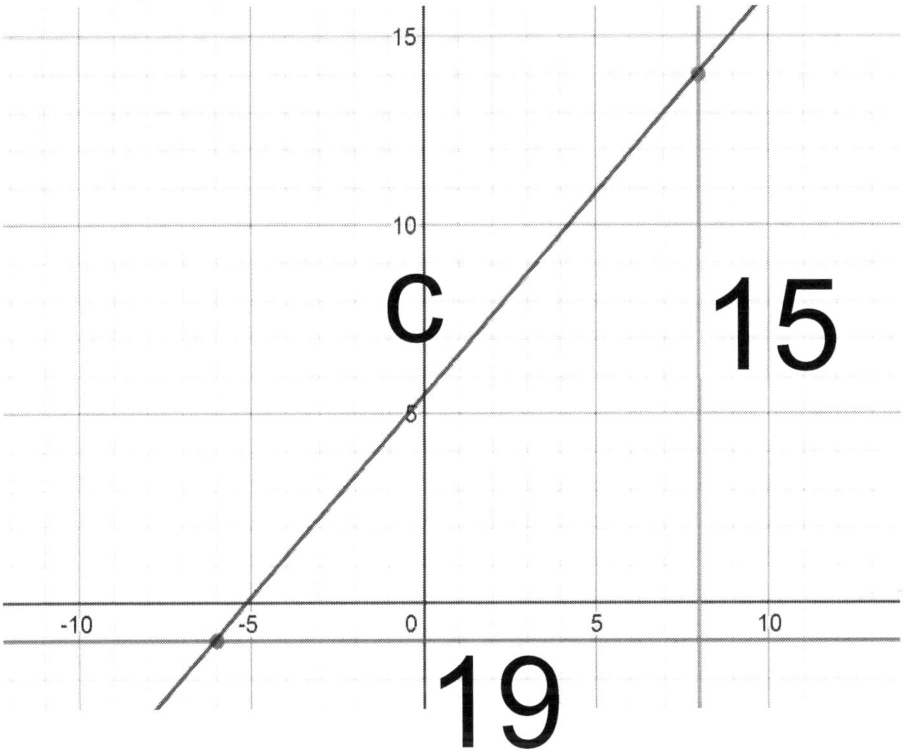

We use the formula to find c:

$$15^2 + 19^2 = c^2$$
$$225 + 361 = c^2$$
$$586 = c^2$$
$$\sqrt{586} = \sqrt{c^2}$$

24.207 is approximately c.

So (-6, -1) and (8, 14) are approximately 24.207 units apart.

Example 2: What is the distance between (-4, 5) and (-1, 10)? Remember to graph your points first:

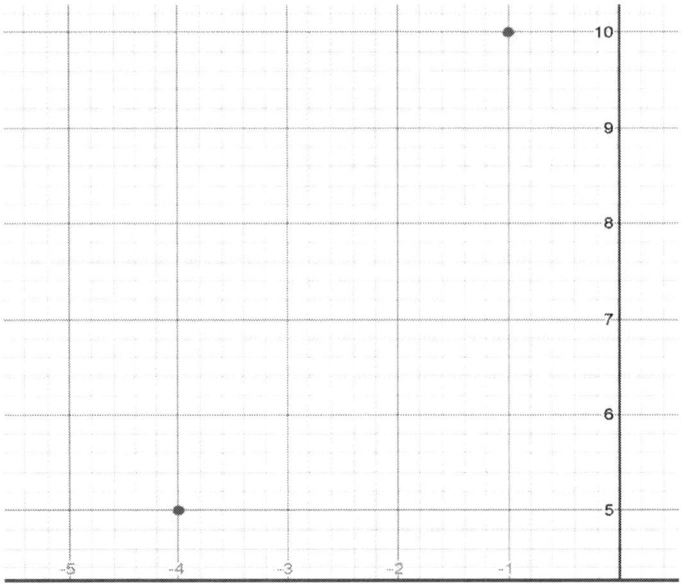

Then, turn your points into a right triangle.

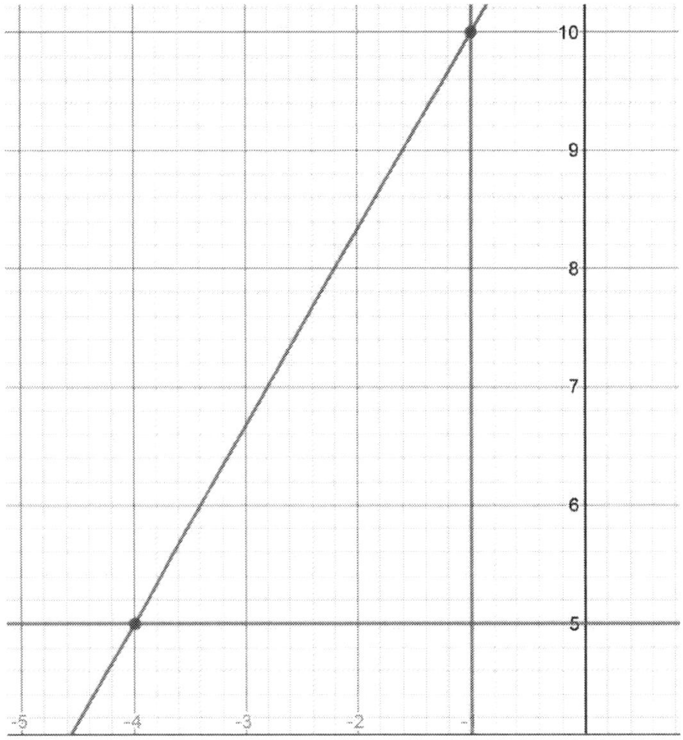

Now, find your values for a, b and c.

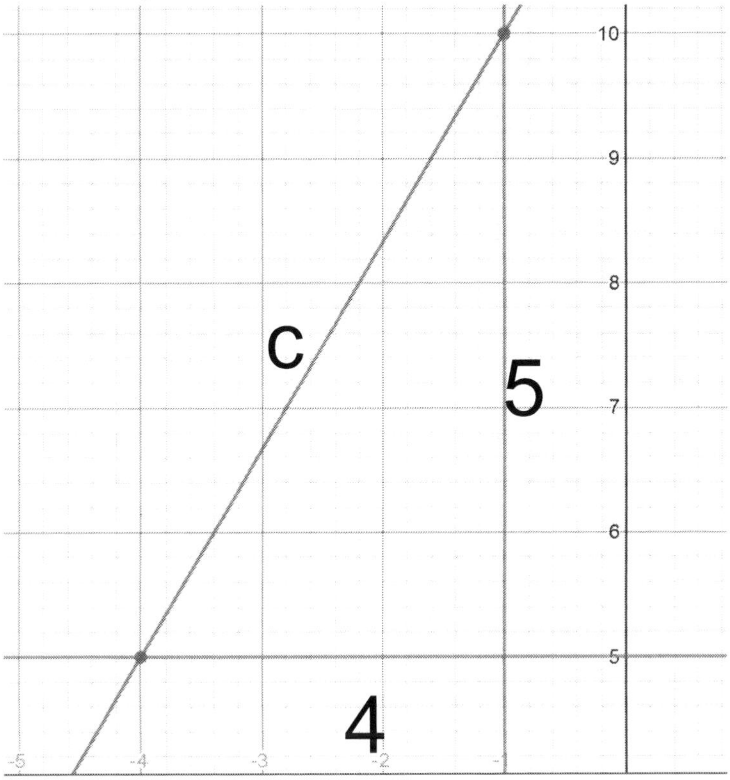

We use the formula to find c:

$$4^2 + 5^2 = c^2$$
$$16 + 25 = c^2$$
$$41 = c^2$$
$$\sqrt{41} = \sqrt{c^2}$$

6.403 is approximately c.
So (-4, 5) and (-1, 10) are approximately 6.403 units apart.

The Distance Formula

Another way to calculate the distance between two points is to use the Distance Formula.

The Distance Formula is a formula you can use by simply plugging in the values you know for x and y.

The Distance Formula is $d = \sqrt{(x_2 - x_1)^2 + (y_2 - y_1)^2}$. To find d, you plug in the values you know for x and y. This formula is especially helpful when you are given two points and unable to graph them.

For example, find the distance between (15, 8) and (2, 20).

You can use the Distance Formula to do this fairly simply.

Let's identify $x_1, x_2, y_1 \text{ and } y_2$. For the points, (15, 8) and (2, 20), we will use the following:

x_1=15

y_1=8

x_2=2

y_2=20

Now, let's plug those values into the formula:

$d = \sqrt{(2 - 15)^2 + (20 - 8)^2}$

Now, we simplify:

$d = \sqrt{(-13)^2 + (12)^2}$

Then, we solve:

$d = \sqrt{169 + 144}$

$d = \sqrt{313}$

$d = 17.7\ units$

Let's review with another set of points: (-4, -9) and (3, -15).

x_1= -4

y_1=-9

x_2= 3

y_2= -15

$d = \sqrt{(3 - -4)^2 + (-15 - -9)^2}$

$d = \sqrt{(7)^2 + (-6)^2}$

$d = \sqrt{49 + 36}$

$d = \sqrt{85}$
$d = 9.2\ units$

Practice problems:

1. What is the distance between (-3, -6) and (1, 3)?
2. What is the distance between (-14, 8) and (4, 4)?
3. What is the distance between (7, 2) and (-8, 9)?
4. What is the distance between (12, -5) and (-11, -7)?
5. What is the distance between (-5, -9) and (-2, -10)?

Answers

1. 9.85 units
2. 18.44 units
3. 16.55 units
4. 23.09 units
5. 3.16 units

Functions

A function is a relationship that relates an input to an output.
Usually functions are written in terms of x.
We use the representation f(x) to represent the idea that the function is related to x.
What this means is as x changes, the value of the function (or the y value) will change.
Functions can be broken down into three parts:

The input, or the value you are starting with.
The relationship or rule that will manipulate the value.
The output, or the value you are ending with.

Generally, the input is represented by x and the output is represented by f(x), which is also the y value.

Look at this table below.
It represents a very basic function.

x	y
2	5
4	7
7	10

Can you guess the relationship between x and y?
As x increases, y increases by 3. So we would say the rule of this function is $y = x + 3$.
An important thing to keep in mind about functions is that each input can only have one output.
In other words, each x value can only calculate one y value.
This can be confusing because sometimes different x values will give you the same y value.
That is ok, but each x value can only calculate one y value.

When we talk about functions, we normally also discuss domain and range.
The domain of the function is all x values that exist for that function.
The range of the function is all y values that exist for that function.
For the table above, the domain of the function is {2, 4, 7} and the range is {5, 7, 10}.
Sometimes the domain or range can be all numbers. This is especially the case when looking at functions that are lines.
Let's practice identifying some functions!

Explanation/examples:
What is the rule of this function?

x	y
2	3
10	15
18	27

To determine the rule of the function, first we need to figure out the relationship between the x and y values. To find this, we can divide the y value by x and see if each set has a similar relationship. In this case, they are all multiplied by 1.5. Each y value is 1.5 times the x value.

Is this a function?

x	y
1	5
1	14
11	15

To determine whether this is a function, we have to determine that each x only has 1 output. In this relationship, x has 2 different outputs so this is not a function.

Is this a function?

x	y
6	15
8	15
10	23

To determine whether this is a function, we have to determine that each x only has 1 output. In this relationship, x has only 1 output so it is a function. Although there are two x's with the same output, because each x has its own output, it is a function.

What is the rule of this function?

x	y
4	15
7	24
9	30

To determine the rule of the function, first we need to figure out the relationship between the x and y values. Each x value is multiplied by 3 and then 3 is added to the result. The rule of this function can be written $f(x)=3x+3$

Practice problems:

1. Circle the relationships below that are functions and determine the rules.

x	y
2	14
2	18
8	22

Rule: ___________

x	y
1	1
4	2
9	3

Rule: ___________

x	y
1	1
2	4
3	9

Rule: ___________

x	y
1/2	2
3/4	3
7/8	7/2

Rule: ___________

x	y
17	21
13	17
17	21

Rule: ___________

Answers.

x	y
2	14
2	18
8	22

Rule: Not a function

x	y
1	1
4	2
9	3

Rule: $\sqrt{x}$

x	y
1	1
2	4
3	9

Rule: x^2

x	y
1/2	2
3/4	3
7/8	7/2

Rule: Multiplied by 4.

x	y
17	21
13	17
17	21

Rule: Not a function

Graphs of Functions

We have looked at functions that are represented by charts but functions can also be represented by graphs.

A function can be graphed when there are values given for the domain and range. Remember, the domain are the x values of the function and the range are the y values of the function.

Graphs can be useful ways to represent the function relationship. We can also use graphs to determine whether a relationship is a function.

Relationships that are functions will pass the vertical line test. When we draw a vertical line through any part of a function, we should only have one point on that line. If we have more than one point on the line, then we say the relationship has failed the vertical line test and is not a function.

Here is a function:

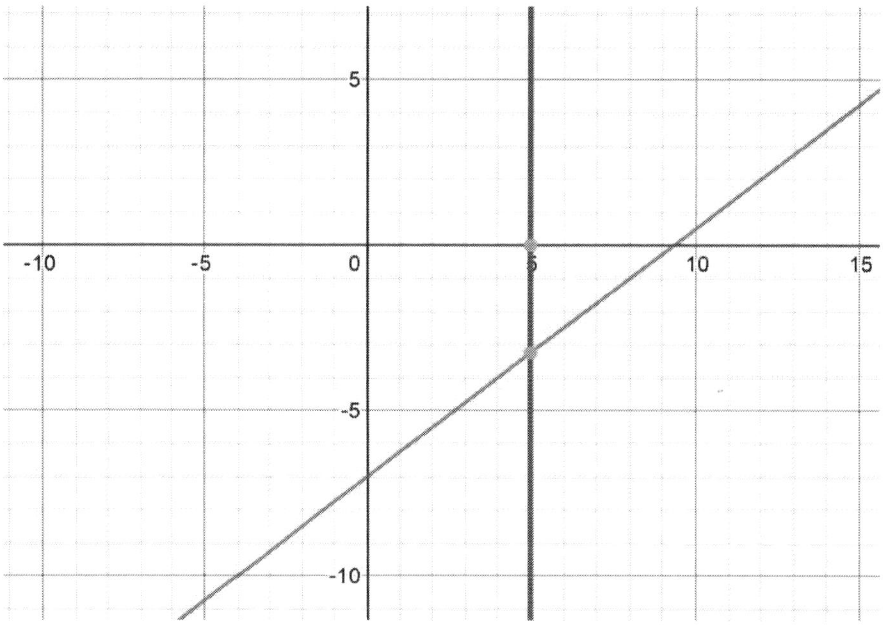

Not a function

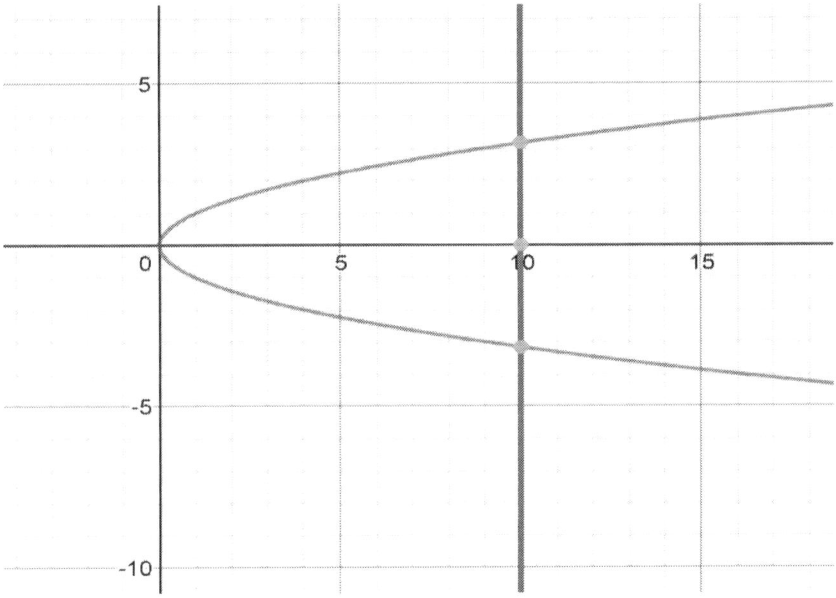

There are different types of functions based on the variable x.
The most common type of function is the linear function.
This function consists of a straight line.
Linear functions are most commonly written in the form y=mx+b, where m is the slope of the line and b is the point where the line intercepts the y-axis.

Here is an example of a linear function.

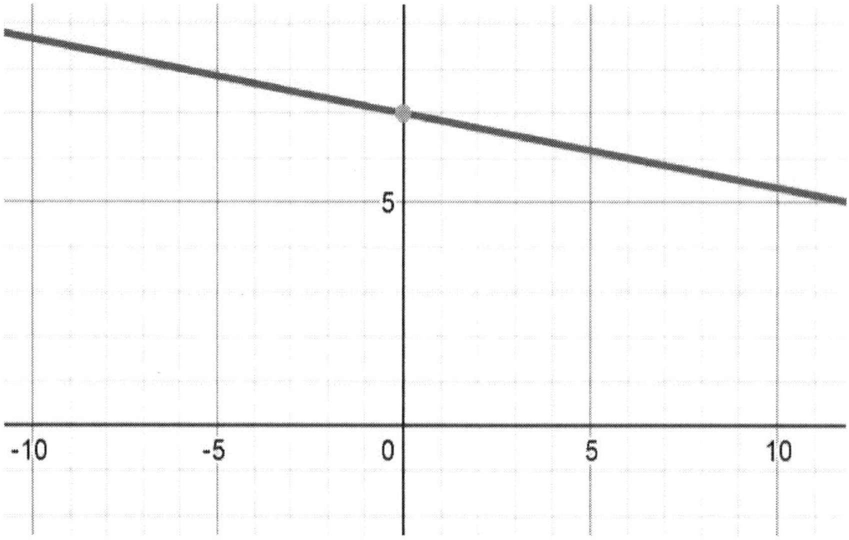

Another type of function is a quadratic function.
A quadratic function occurs when x is raised to a second order power.
Quadratic functions do not involve straight lines but curves.
A quadratic function is usually shaped like a parabola with a vertex, which is its highest or lowest point. The vertex also acts as a point of symmetry for the quadratic function.
Here is an example of a quadratic function.

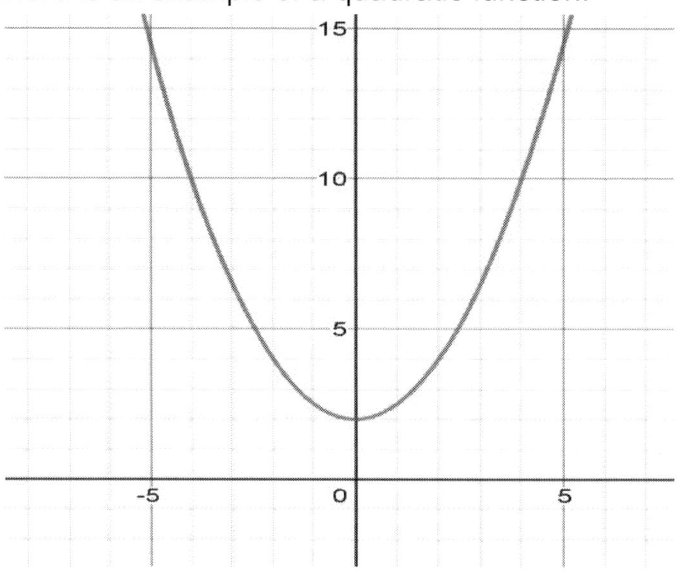

Explanation/examples:

Use the vertical line test to determine if the graph below is a function.

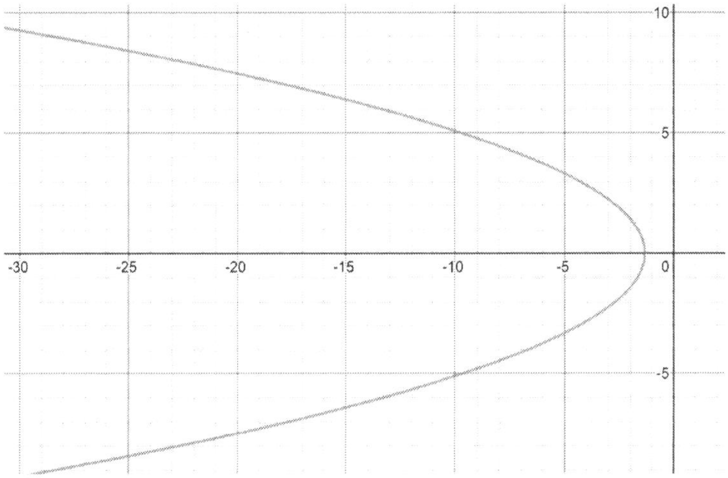

Pick a point on the graph to draw a straight line.

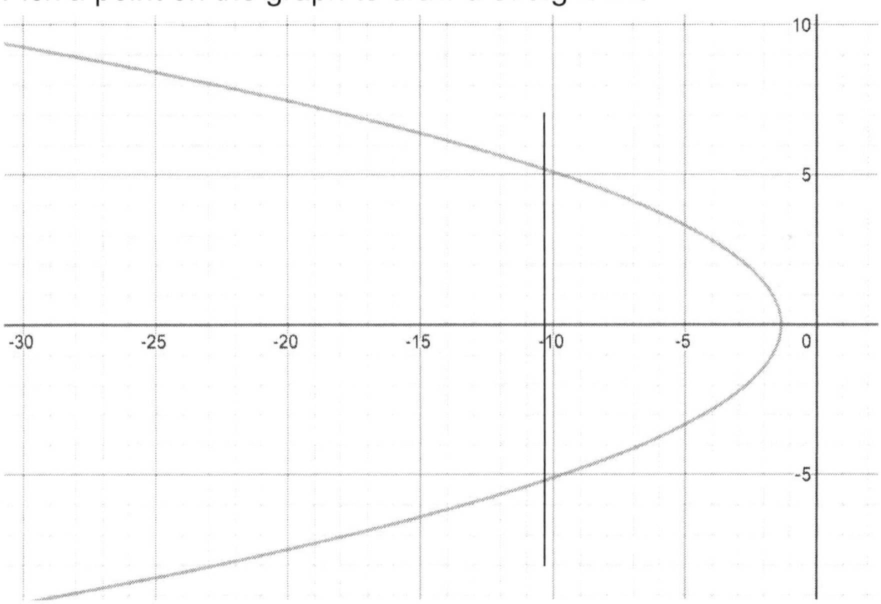

Because the line goes through two points of the graph, this is not a function.

Identify this function.

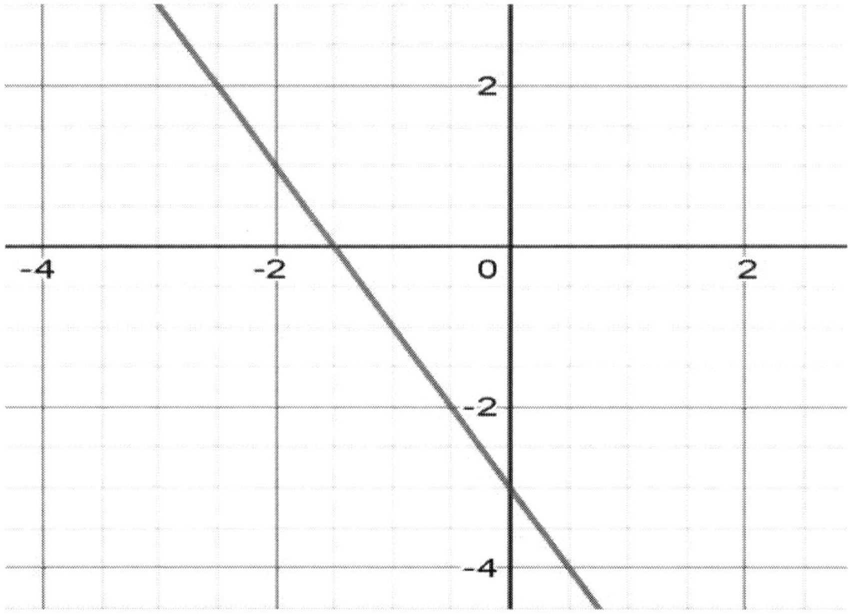

This graph is a linear function because it follows the form y=mx+b and it is a straight line.

Practice problems:

1. Use the vertical line test to determine if the graph below is a function.

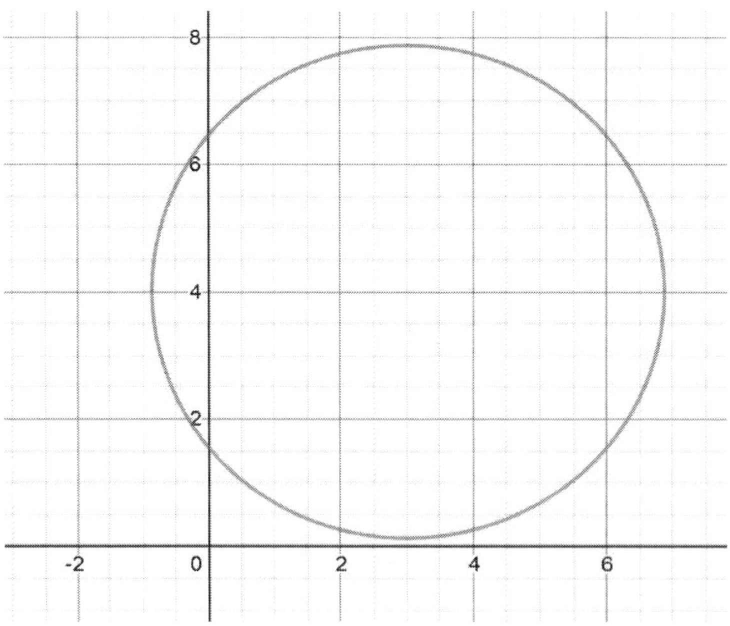

2. Use the vertical line test to determine if the graph below is a function.

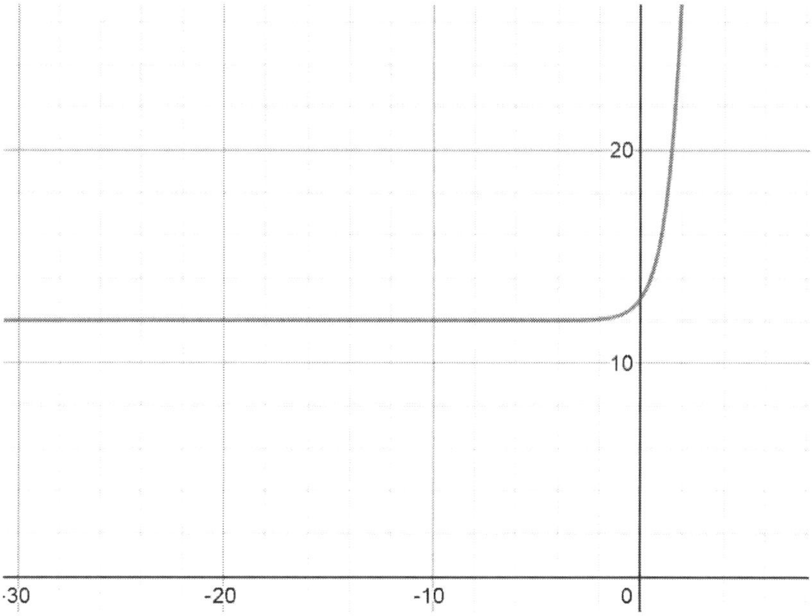

3. What type of function is this?

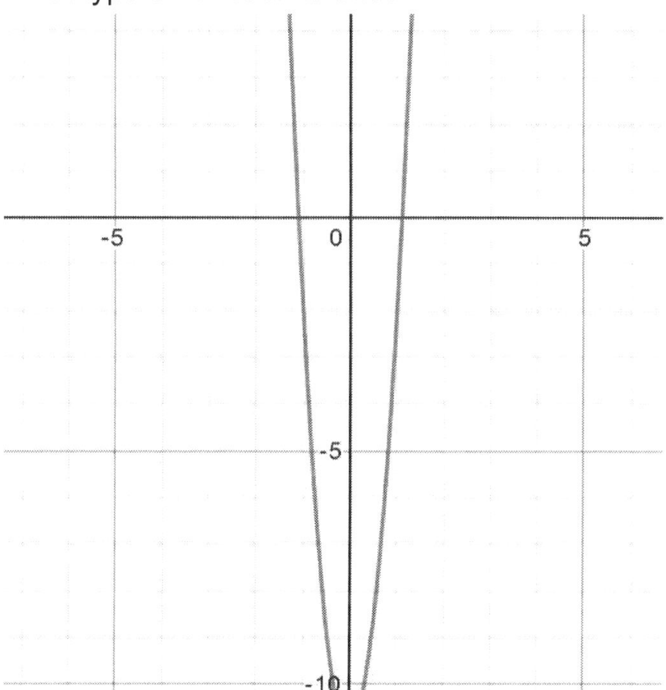

4. What type of function is this?

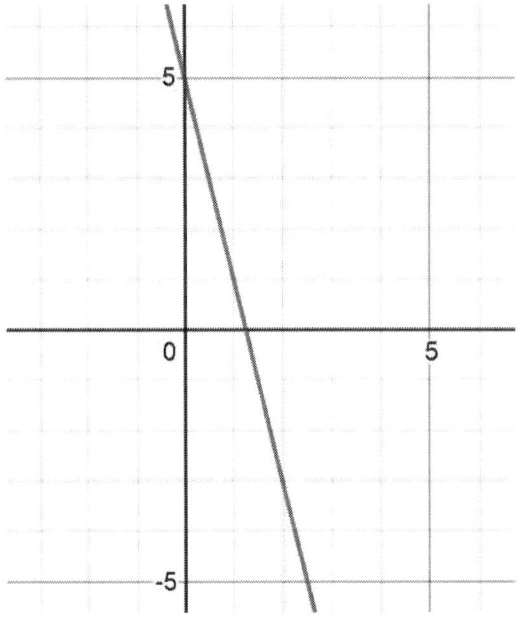

5. What type of function is this?

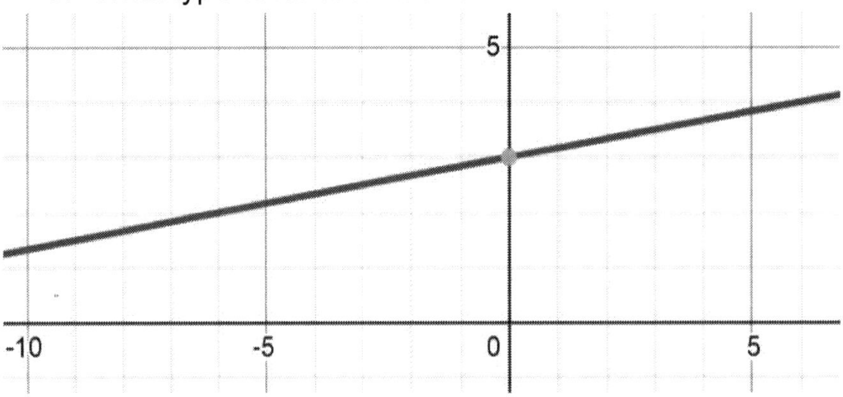

Answers:

1. Not a function
2. Function
3. Quadratic
4. Linear
5. Linear

Angles and Arcs

Angle in Standard Position

Trigonometry relies on an understanding of the unit circle. The unit circle is a circle with a radius of 1 that has its center at the origin or (0, 0). That means at any point on the circle, the radius of the circle is 1 unit long. We will return to this unit circle later to help us explore triangles but for now, let's examine how it helps us understand angles and circles.

An angle is said to be in standard position if its vertex is located at (0,0). When an angle is in standard position, one side is made up of the x-axis. This side is known as the initial side. The other side of the angle is known as the terminal side. To find the measurement of the angle, you calculate the rotation from the initial side to the terminal side.

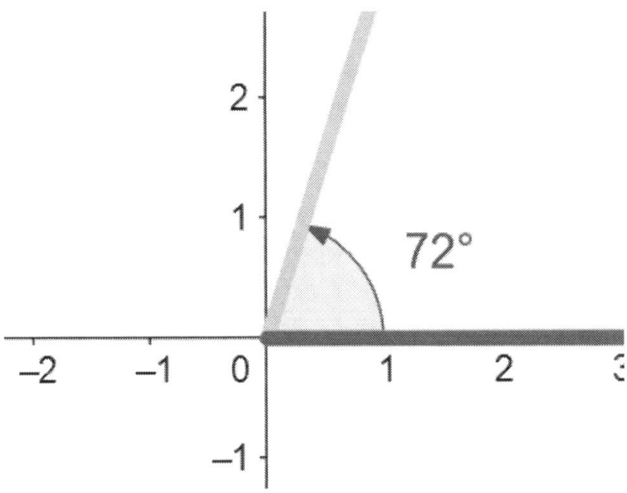

To determine if an angle is in standard position, you do the following:

1. Determine whether the vertex of the angle intersects at (0,0).
2. Be sure one side of the angle is overlapping the x axis.
3. Review the angle to be sure it shows measurement when looking at the angle counter clockwise.

Let's draw an angle in standard position.
Draw the angle 17°.
First, create an axis below.

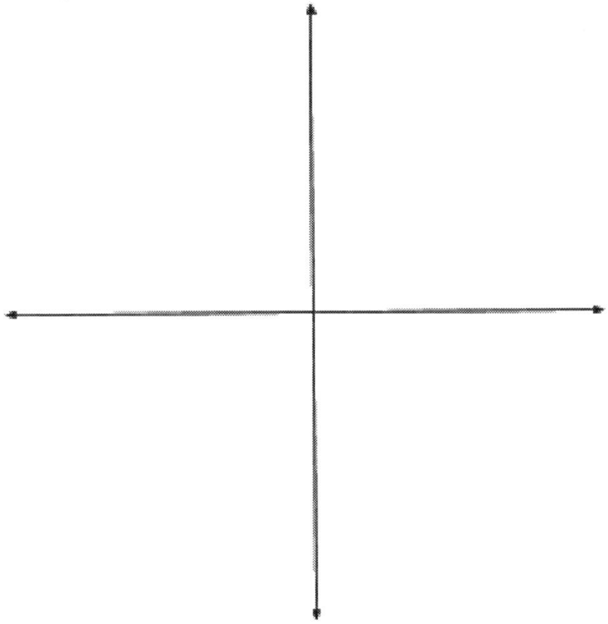

Next, draw a line on the x axis, with the vertex at (0, 0).

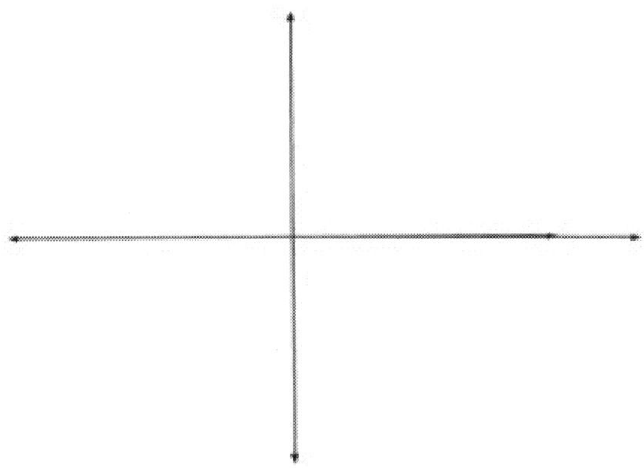

Finally, use a protractor to draw the second ray so it represents a measurement of 17°.

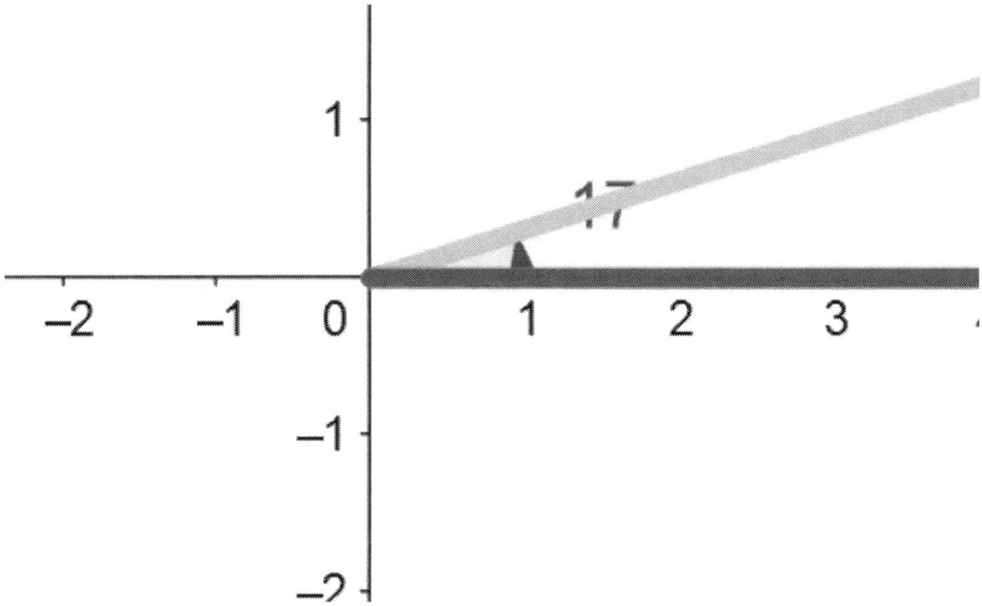

Draw the following angles in standard position.

1. 42°
2. 60°
3. 75°
4. 8°
5. 124°

Answers

1.

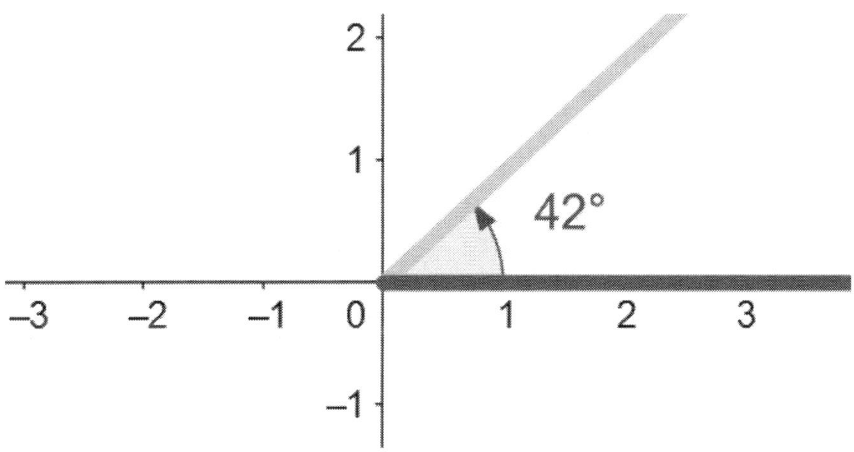

2.

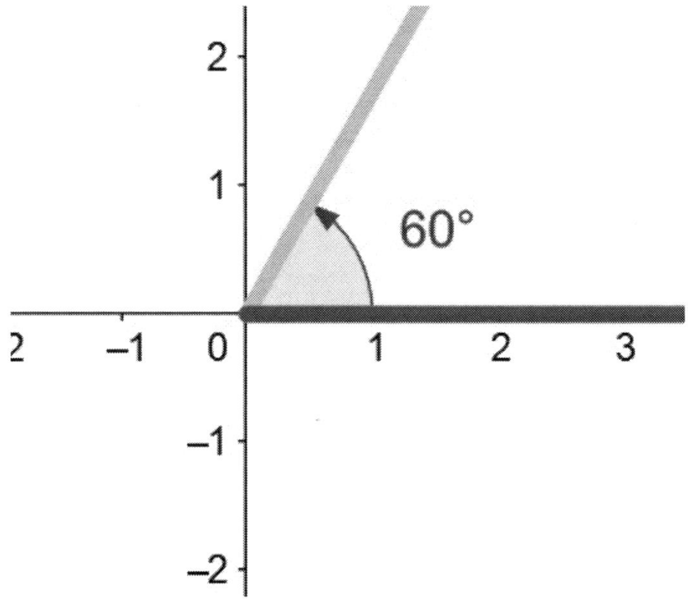

3.

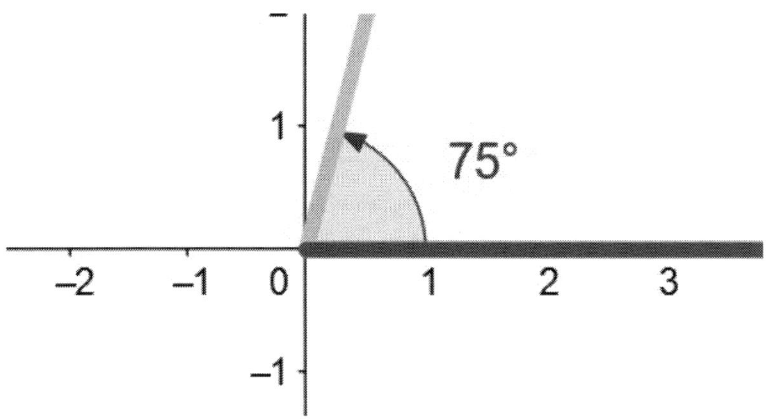

4.

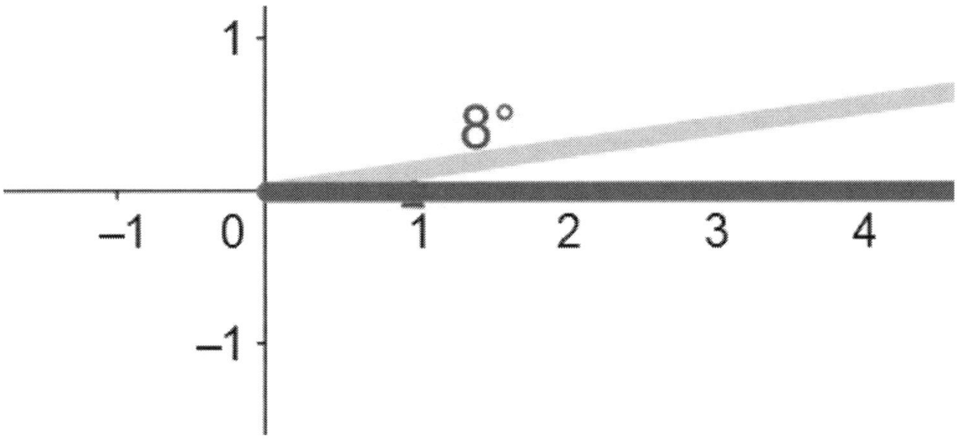

5.

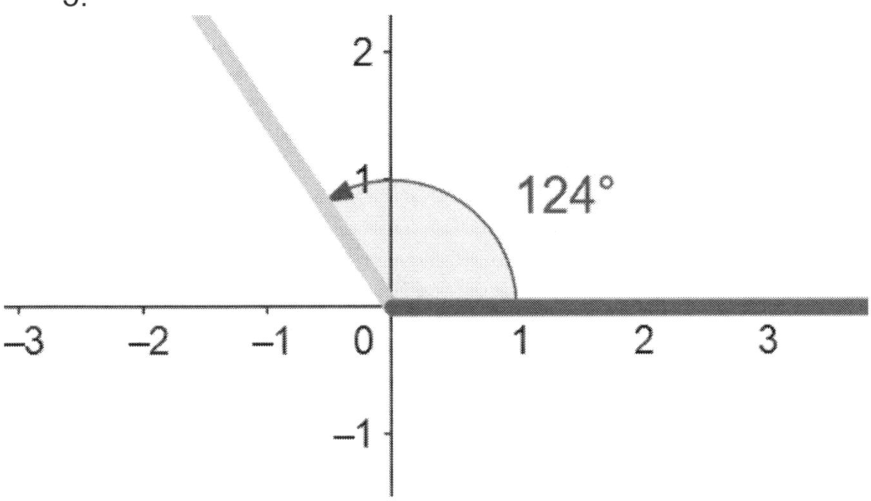

Angle Measure

To measure an angle, you may use a protractor. You should line up the bottom of the protractor with the x axis and then use the terminal side to find the measurement.

Let's practice!

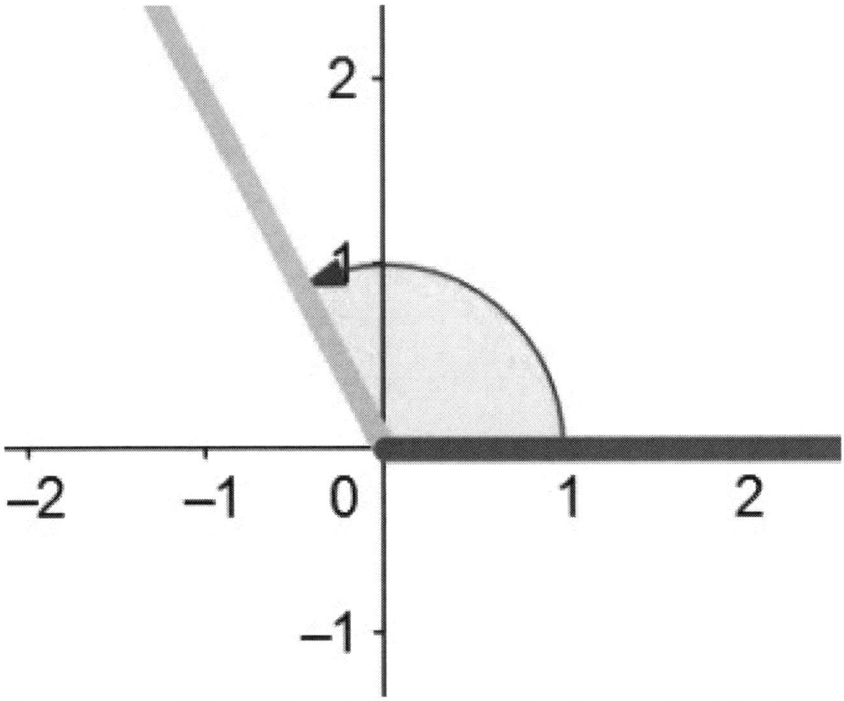

To measure this angle, place your protractor on the blue line. Look at where the yellow line crosses your protractor. You should be able to calculate the measurement of 118°.

Practice Measuring the Angles Below:

1. 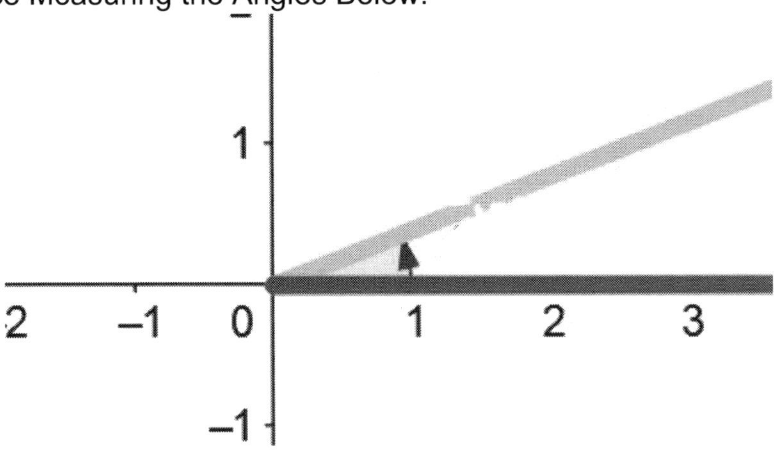

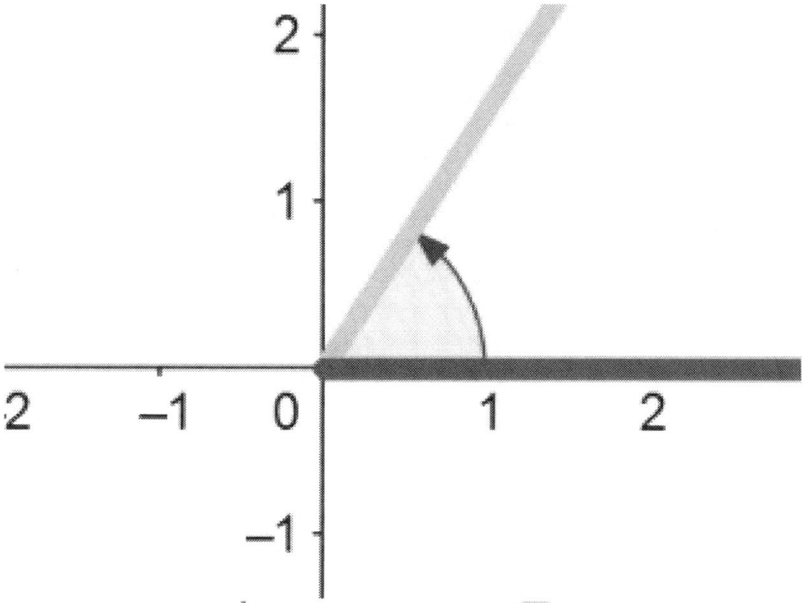

2.

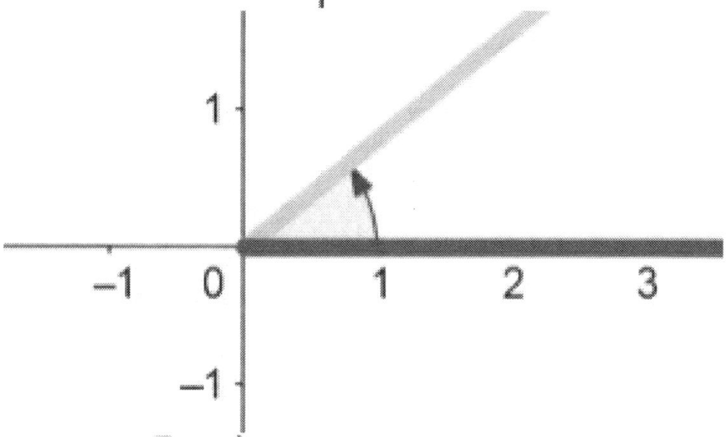

3.

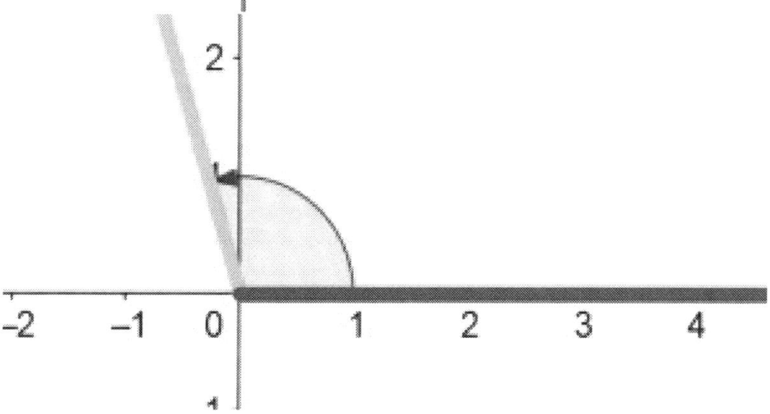

4.

5. 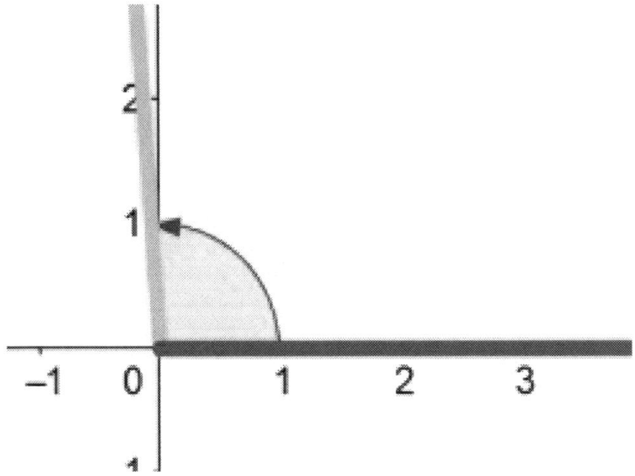

Answers

1. 21°
2. 57°
3. 38°
4. 106°
5. 94°

Converting Between Radians and Degrees

Many students have experience manipulating angles with degrees but a new term to many trigonometry students is radians. Radians are calculated using the unit circle. If the radius is 1 in the unit circle, then we can calculate the circumference of the unit circle as 2 π This makes one revolution of the unit circle to be measured as 2 π.

This means 360°= 2 π radians. To make it easier when converting degrees, we can simplify to 180°= 1 π radians.

Let's look at the example:

How would you convert π/4 radians to degrees?

The easiest way to convert is to set up a proportion:

π/4 * 180/ π = 180/4 = 45°

Practice converting angles below.

1. Convert 60° to radians.
2. Convert 0.2 radians to degrees.
3. Convert 85° to radians.
4. Convert 130° to radians.

5. Convert 1.8 radians to degrees.

Answers

1. 1.04
2. 11.4°
3. 1.48
4. 2.26
5. 103°

Coterminal Angles

Angles may also have coterminal angles. Coterminal angles share the same terminal side.

In the example below, the angles are coterminal as they share the yellow line in quadrant II.

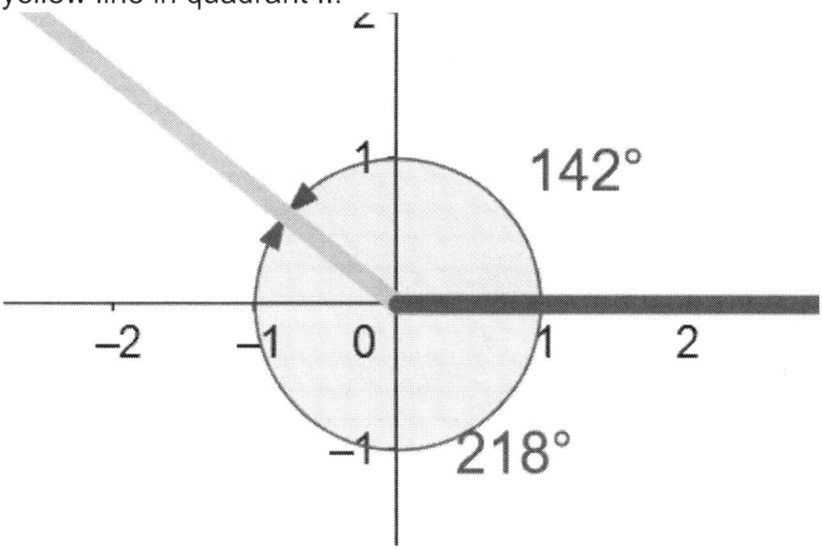

Coterminal angles add up to 360°. To find the measurement of a coterminal angle, you simply subtract the given angle from 360°. For example, for the first angle, we are given a measurement of 39°. What would be the coterminal angle for the given angle? 360-39=321. The coterminal angle would be 321°.

Practice finding the coterminal angles for the given angles below.

1. 174°
2. 294°
3. 87°
4. 92°
5. 307°

Answers:

1. 186°
2. 66°
3. 273°
4. 268°
5. 53°

Arc Length and Area of a Sector

We can use our knowledge of the angles of a unit circle to calculate its arc length and area of a sector or part of a circle based on our knowledge of radians.

A whole circle has an area of πr^2. We also know its angle in radians is 2 π

A sector is a part of a circle. Its area will still be calculated by the formula πr^2 but its angle is 2π.

We can simplify this formula to πr^2, which is in radians.

When we are given the angles in degrees, we should use the formula 360 πr^2,.

Similarly, we have calculated the arc length by the formula l = rθ where θ is in radians

When we are given the angle in degrees, we should use the formula

Let's practice! Find the area and arc length of a sector that measures 128° and has a radius of 3.

A=½ * 32π /45 * 32
A= 10.05

L=3 * 32π /45
L= 6.702

Problems

1. Find the area of a sector that measures 71° that has a radius of 2.
2. Find the area of a sector that measures 7° that has a radius of 5.
3. Find the arc length of a sector that measures 159° that has a radius of 11.
4. Find the area of a sector that measures 35° that has a radius of 0.5.

5. Find the arc length of a sector that measures 4° that has a radius of 8.
6. Find the arc length of a sector that measures 28° that has a radius of 4.

Answers

1. 2.48
2. 1.527
3. 30.5
4. 0.07
5. 0.559
6. 1.955

Angles created using

https://www.geogebra.org/m/nC98H4NH

Right Triangle Trigonometry

Section 1: Pythagorean Theorem

We have already looked at the Pythagorean Theorem but let's review again. The Pythagorean Theorem is a way to calculate information about right triangles.

Here is a right triangle.

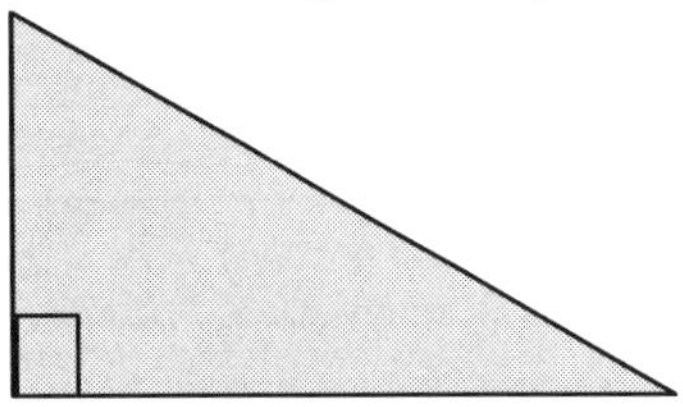

We know it's a right triangle because there is a square in the corner. That square indicates the angle's measurement is 90°. Triangles with right angles have a special way to find the length of the sides.

To find the length of the side of a right triangle, you can use the Pythagorean Theorem. Remember, the Pythagorean Theorem is the formula: $a^2 + b^2 = c^2$.

A and B and C all stand for specific sides. C is always the hypotenuse or the side opposite the right angle. A and B are always the sides around the right angle.

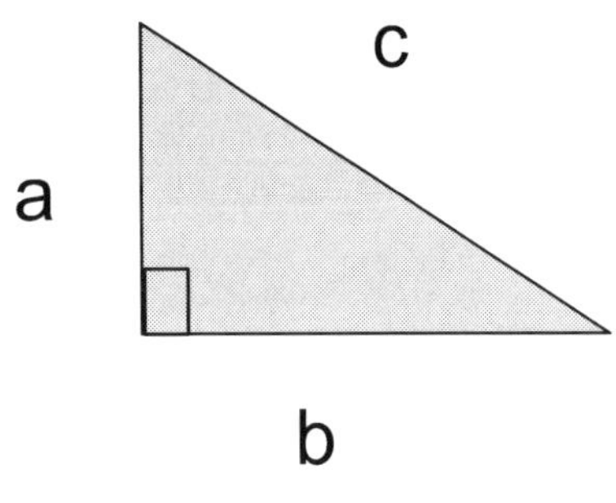

If we are given information about two sides, we can calculate the third side based on the Pythagorean Theorem. It doesn't matter which two sides we are given information on.

Let's look at a sample problem.

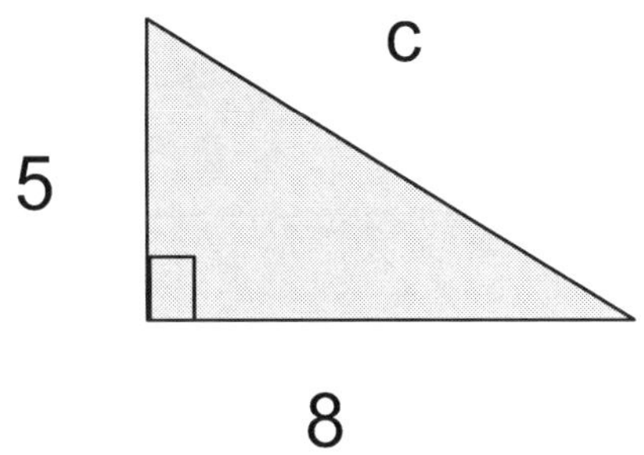

First, let's determine what we know.
We know a=5 and b=8 and we are trying to calculate c.
We can use the Pythagorean Theorem to calculate c.

$$a^2 + b^2 = c^2$$
$$5^2 + 8^2 = c^2$$
$$89 = c^2$$
$$\sqrt{89} = \sqrt{c^2}$$
$$9.43 = c$$

Let's practice.

1. Find the measurement of the missing side.

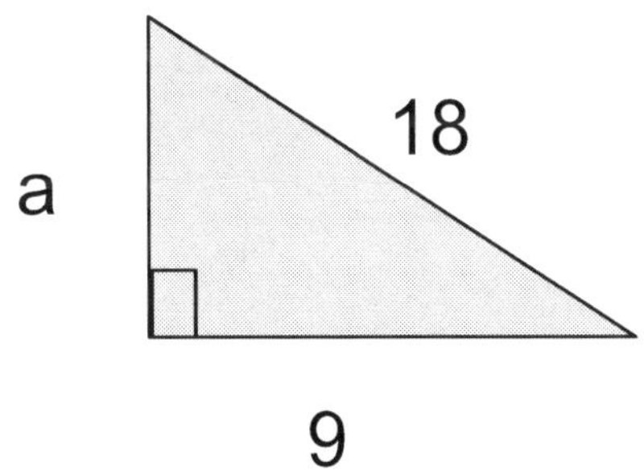

2. Find the measurement of the missing side.

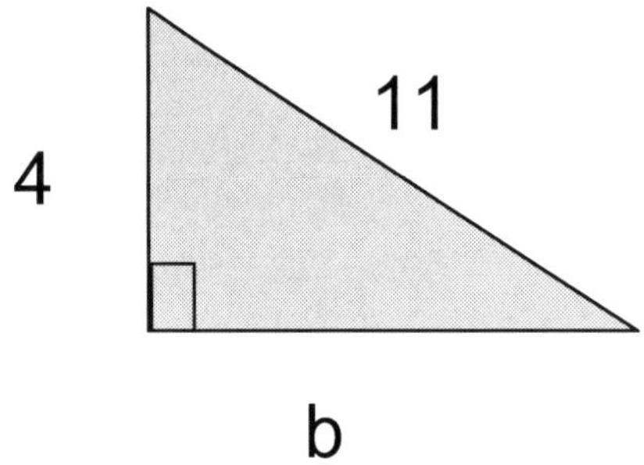

3. Find the measurement of the missing side.

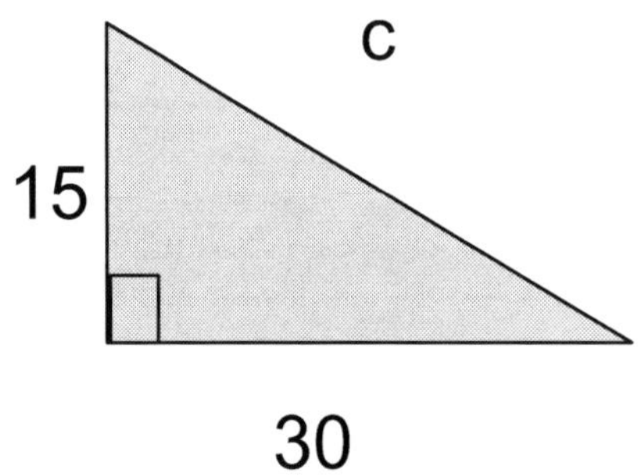

4. Find the measurement of the missing side.

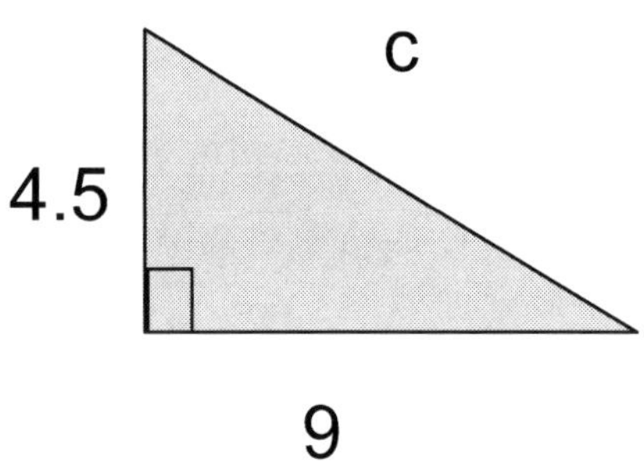

5. Find the measurement of the missing side.

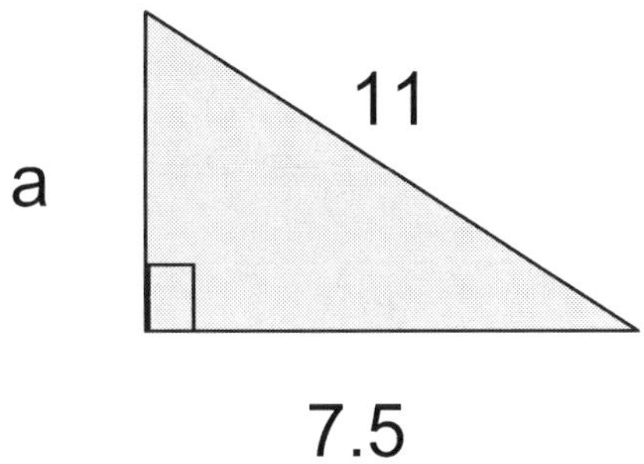

Answers.

1. 15.588
2. 10.247
3. 33.541
4. 10.062
5. 8.047

Section 2: The Six Trigonometric Functions

The study of trigonometry applies information about right angles and the Pythagorean further. Let's look at a few terms first.

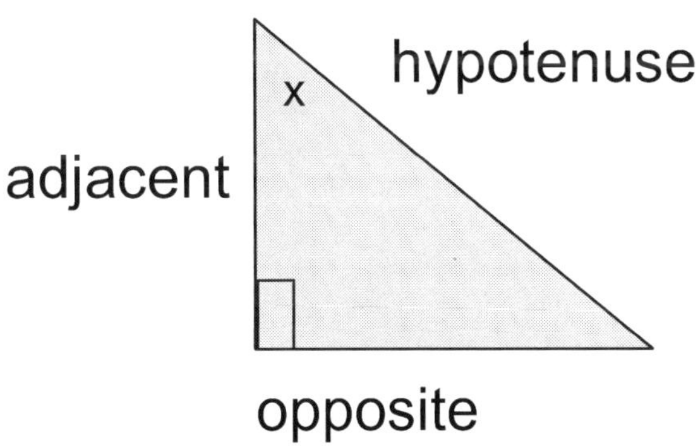

The angle that we are interested in is above the right angle, marked with an x. We call it *theta* and it is represented by the symbol: θ in calculations.

We can use theta to define the sides of a right triangle.

Hypotenuse: The side opposite the right angle.

Opposite: The side opposite the theta angle.

Adjacent: The side connecting both the right angle and the theta angle.

The trigonometric functions are designed as follows:

1. Sine (angle): measure of the opposite side of theta divided by measure of hypotenuse
2. Cosine (angle): measure of adjacent side of theta divided by measure of hypotenuse
3. Tangent (angle): measure of the opposite side of theta divided by measure of adjacent side of theta
4. Secant (angle): measure of hypotenuse divided by measure of adjacent side of theta
5. Cosecant (angle): measure of hypotenuse divided by measure of the opposite side of theta
6. Cotangent (angle): measure of adjacent side of theta divided by measure of the opposite side of theta

The last three functions are used less often and can be calculated as 1 over sine, 1 over cosine and 1 over tangent.

We can use a calculator and information we know to calculate what we are trying to find.

Let's look at a sample problem.

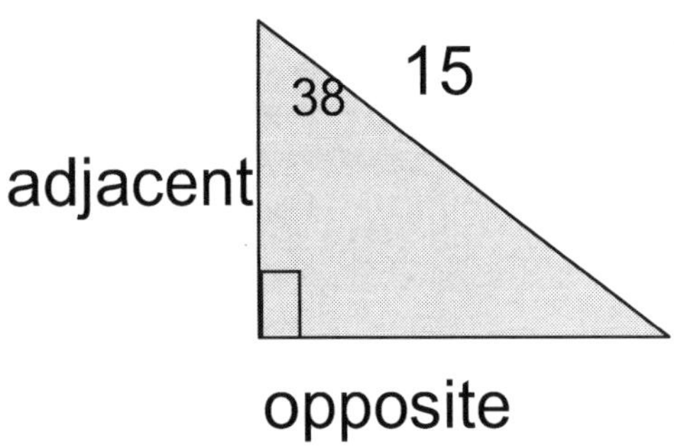

If we know the measure of theta and the measure of the hypotenuse, we can use that information and a calculator to calculate a variety of functions.

Let's calculate the adjacent side.

Cosine (angle): measure of adjacent side of theta divided by measure of hypotenuse

Cosine (38)= x/15 Use a calculator to calculate cosine 38.

0.95507=x/15 Multiply both sides by 15.

x= 14.326

The adjacent side is 14.326 units long.

Let's practice.

1. Use the information below to find the length of the opposite side.

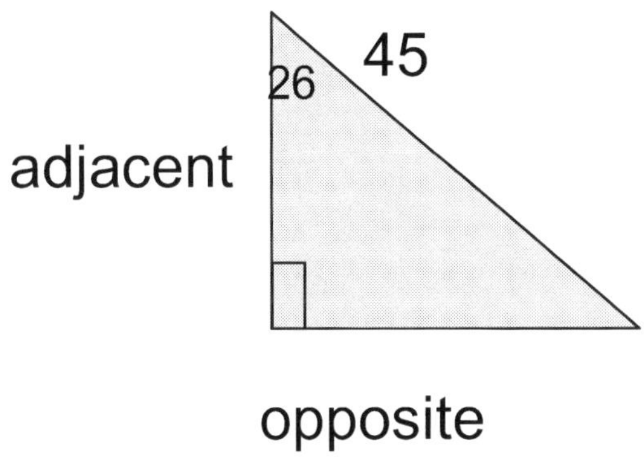

2. Use the information below to find the length of the adjacent side.

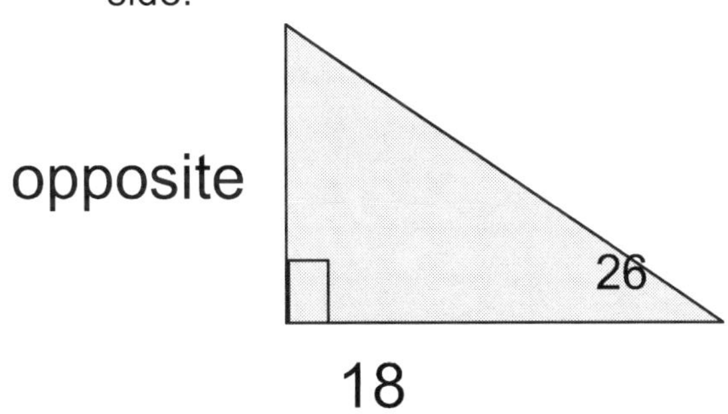

3. Use the information below to find the length of the hypotenuse.
Cosine (angle): measure of adjacent side of theta divided by measure of hypotenuse

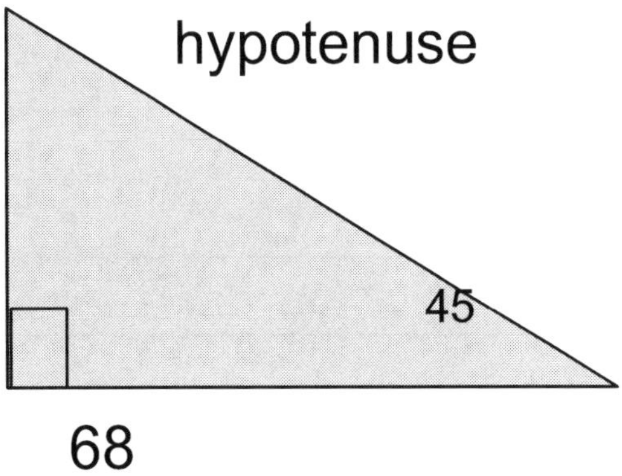

Answers.

1. 34.315
2. 21.218
3. 129.444

Section 3: Two Special Right Triangles

There are two types of right triangles that are easy to solve in geometry. Both are based on their angle measures.
The first triangle has 2 angles that measure 45° and one angle that measures 90°.
Because it has two angles that are the same, we can also make the conclusion that it has two sides that are the same.

We can use that fact and the Pythagorean formula to give us a formula to find the hypotenuse when we know the two sides. If we know the two sides are the same, measurement a, we can find the hypotenuse by using the formula $hypotenuse = a\sqrt{2}$.

The second triangle has angles that measure 30°, 60°, and 90°. Because these angles relate to each other, we can find all three sides of the triangle based on its smallest side, a, which is across from the 30° angle.

If the smallest side is a, then the second leg (across from the 60°) is $a\sqrt{3}$ and the hypotenuse is $2a$.

Let's look at a sample problem.

You are given the measurement 15 feet for a right triangle with 2, 45° angles. Using the formula, we can calculate the hypotenuse as $15\sqrt{2}$ = 21.213

You are given the measurement of 8 inches for a 30, 60, 90 triangle side a. What is the second leg and the hypotenuse?

The second leg is $a\sqrt{3}$=$8\sqrt{3}$ =13.856 inches

The hypotenuse is $2a = 2 \times 8 = 16$ inches

Let's practice.

1. Find the hypotenuse of an isosceles right triangle with a side length of 22.
2. Find the side of an isosceles right triangle with a hypotenuse of 7.5.
3. Find the medium side of a 30, 60, 90 right triangle with a small side of 17.
4. Find the largest side of a 30, 60, 90 right triangle with a small side of 61.

Answers

1. 31.113
2. 5.303
3. 29.445
4. 122

Section 4: Applications and Solving Right Triangles

One of the reasons we study trigonometry is to help us determine unknown information. We can look at the information we know and often use trigonometric functions to help us find information that we don't know, especially in real world situations.

One common component of trigonometry is angle of elevation or angle of depression. When you are looking at an object, the angle of elevation or angle of depression is the angle from your line of sight up or down to the object.

You can use your knowledge of trigonometry to calculate how tall an object is if you know how far away you are from the object.

Let's look at a sample problem.

You are looking at a tree that is 48 feet away from you. The angle of elevation of your viewpoint is 25°. How tall is the tree?

Tan (26°)= height/48

Solve for the height.

So, height= 48 * tan (26°)

height= 56.580 feet

Let's practice.

1. You calculate a tree to be 120 feet tall as you view it from a 45° degree angle. How far away are you?
2. You are looking at a tree that is 12 feet away from you. The angle of elevation of your viewpoint is 48°. How tall is the tree?
3. You are looking at a tree that is 81 feet away from you. The angle of elevation of your viewpoint is 60°. How tall is the tree?
4. You calculate a tree to be 32 feet tall as you view it from a 35° degree angle. How far away are you?

Answers.

1. 74.084 feet
2. 14.402 feet
3. 25.923 feet
4. 67.537 feet

Defining Trigonometric Functions Using the Unit Circle

We've looked at trigonometric functions using right angles, let's look at how they are related to the unit circle. Remember, a unit circle is a circle with its center at the origin and a radius of 1. Remember, we use radians to measure the unit circle. A radian is the measure of a central angle that intersects the arc equal in length to the radius of the circle. A full circle is equal to 2 radians. That means 1 radian=180.

We can use the unit circle to extend sine and cosine to all real numbers.

Starting at (1,0), we move along the unit circle counterclockwise until we form the angle we are looking for.

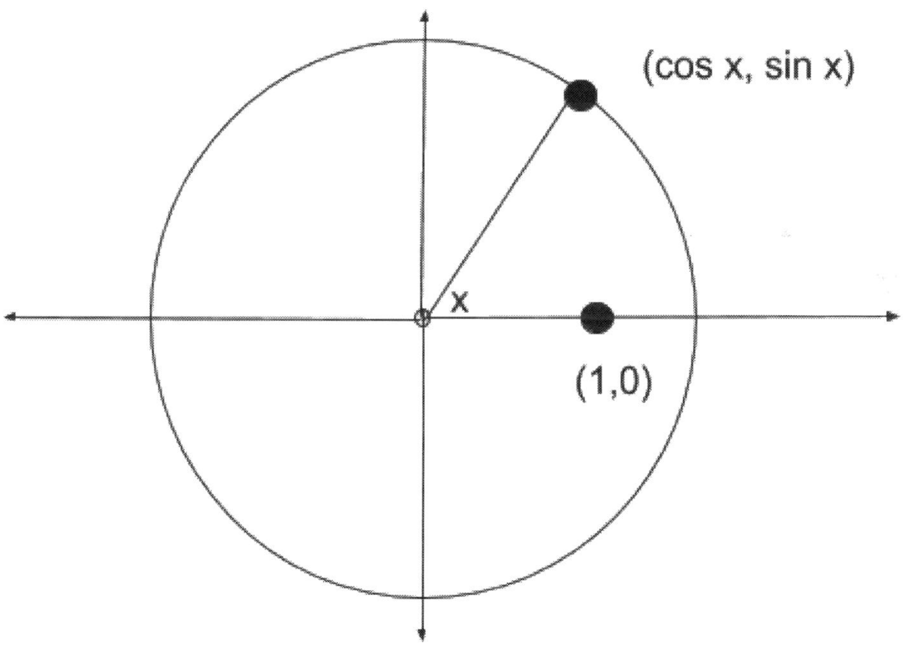

The angle that is formed is indicated by x .

Sin is equal to the y-coordinate of your point and cos is equal to the x-coordinate.

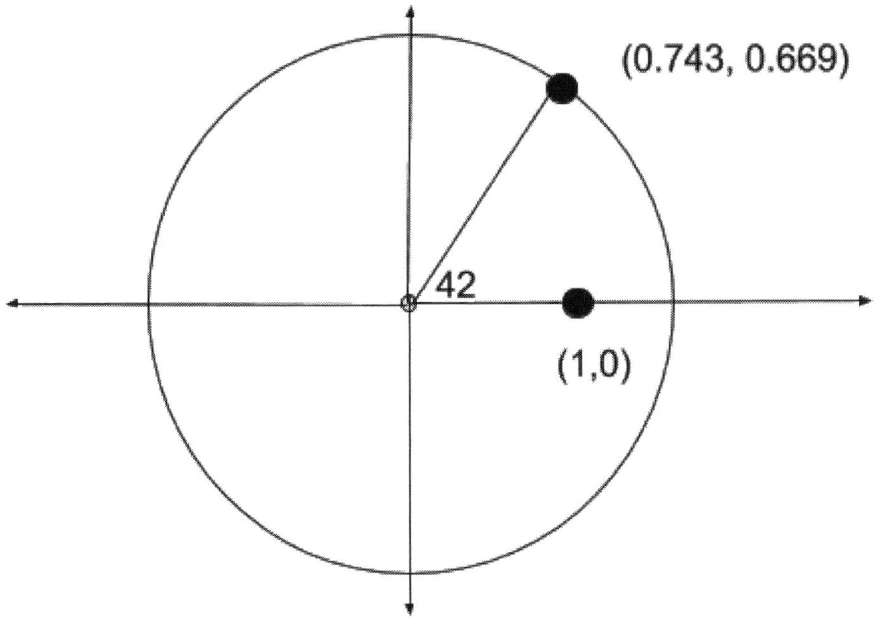

The angle that is formed is indicated by x.

In the illustration above, x =42.

If x =42, sin (42°)= 0.669 and cos (42°)= 0.743

Sin is equal to the y-coordinate of your point and cos is equal to the x-coordinate.

Let's practice!

1. What is the sine of the angle below?

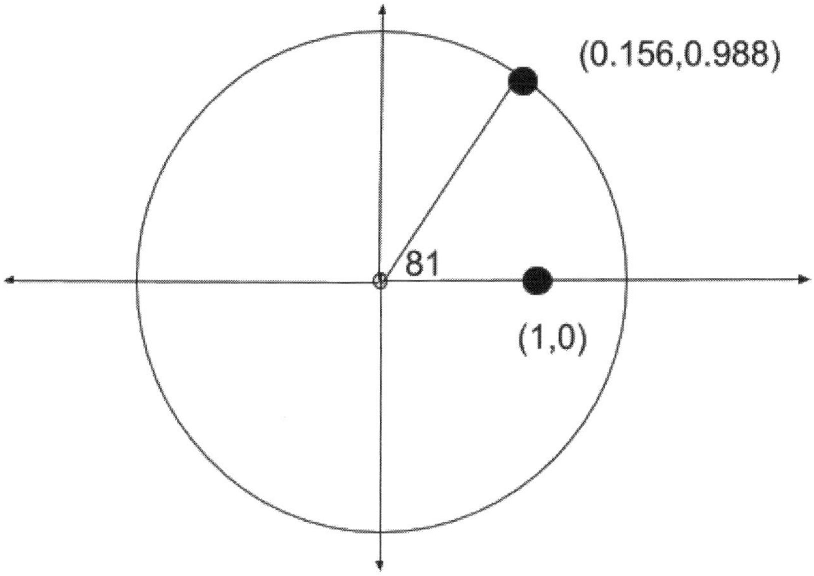

2. What is the cosine of the angle below?

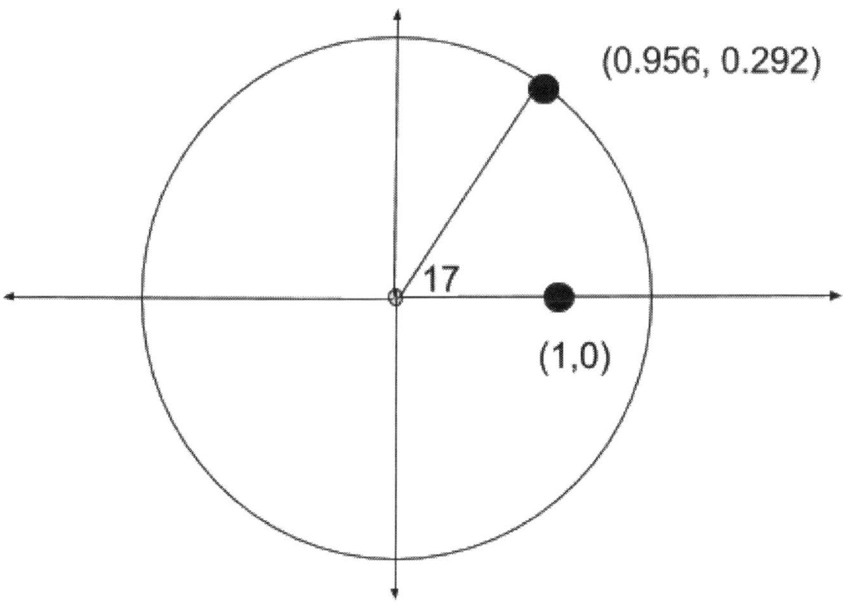

3. What is the cosine of the angle below?

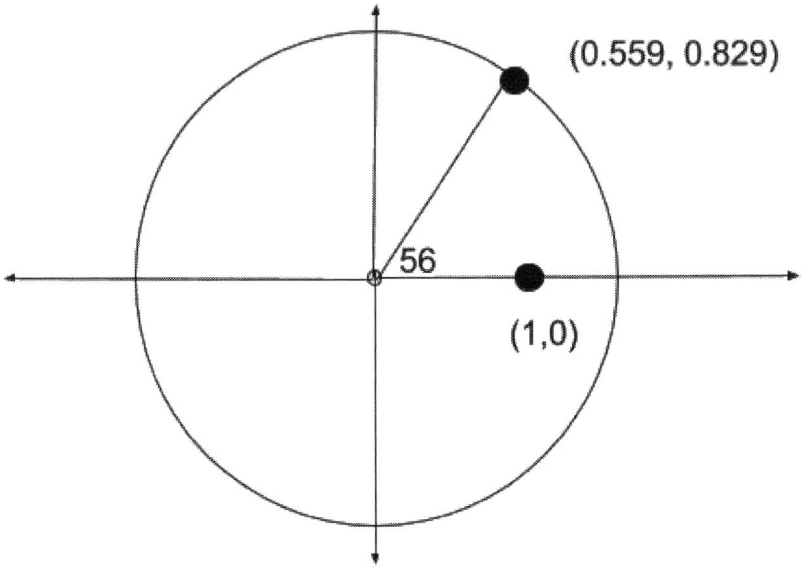

4. What is the sine of the angle below?

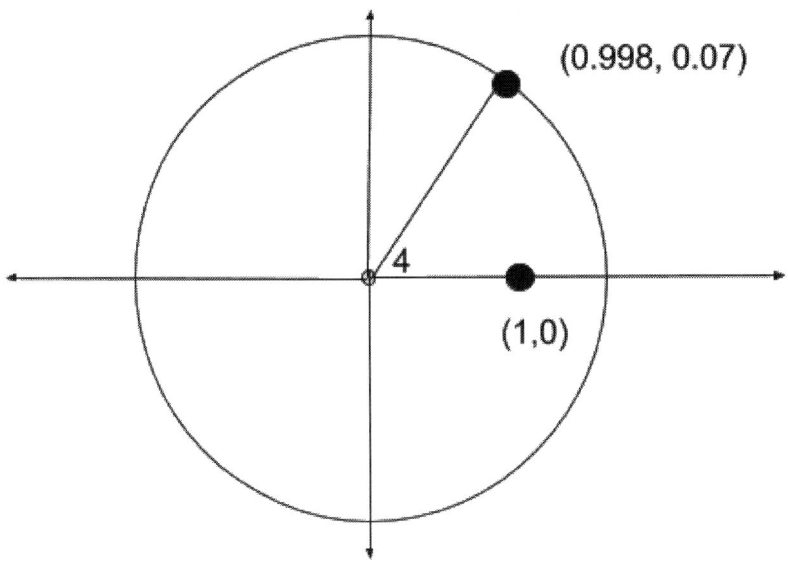

5. What is the cosine of the angle below?

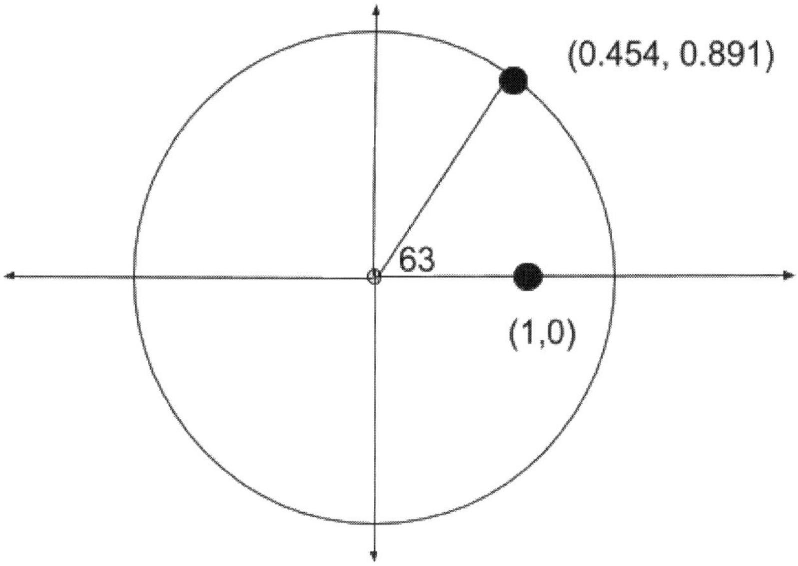

Answers:

1. 0.988
2. 0.956
3. 0.559
4. 0.07
5. 0.454

Pythagorean Identities

In mathematics, we have rules which are called identities.

These are equations that are always true.

There are a few trigonometric identities that can help you find more information. Let's review the unit circle again:

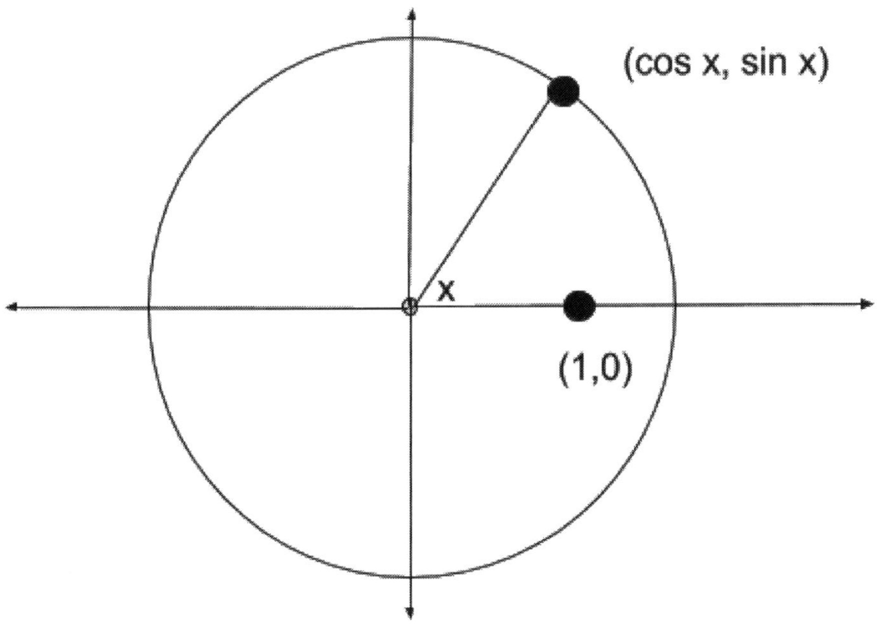

Using the Pythagorean Theorem, we can calculate:

sin 2+cos 2=1

This works for every value of sine.

We can use this identity to create two other identities:

First by dividing each term by cos 2: sin 2cos 2+cos 2cos 2=1cos 2

Using our knowledge of trigonometric functions, we can simplify the identity to

tan 2+1=sec 2

Then we divide each term by cos 2:sin 2sin 2+cos 2 sin2=1sin 2

Using our knowledge of trigonometric functions, we can simplify the identity to

1+cot 2=csc 2

If is located in Quadrant 1, and sin is 1219, what is the cosine?

We can use the Pythagorean identity to calculate the cosine.

12192+cos2= 1

12192-12192+cos2= 1-12192

cos2=0.3988888

(cos2)=0.3988888

cos =0.6315

Let's practice!

1. If is located in Quadrant 1, and cosine is 0.0698, what is the sine?
2. If is located in Quadrant 1, and sine is 0.8746, what is the cosine?
3. If is located in Quadrant 1, and cotangent is 4.705, what is the cosecant?
4. If is located in Quadrant 1, and secant is 1.079, what is the tangent?
5. If is located in Quadrant 1, and sine is 0.8387, what is the cosine?

Answer:

1. 0.9976
2. 0.4848
3. 4.808
4. 0.4040
5. 0.5446

Reference Angles

A reference angle is the positive acute angle that can represent any angle. It must always be less than 90°.

Any angle on the coordinate plane has a reference angle. The reference angle is always the smallest angle from where the angle ends with the x-axis. The reference angle is always defined in terms using the x-axis.

The reference angle depends on the quadrant.

In quadrant I, all angles are acute, so the reference angle is always the angle itself.

In quadrant II, the reference angle can be found using the formula 180-x.

In quadrant III, the reference angle can be found using the formula x-180.

In quadrant IV, the reference angle can be found using the formula 360-x.

Angles in quadrant 1 have measurements between 0° and 90°.

Angles in quadrant 2 have measurements between 90° and 180°.

Angles in quadrant 3 have measurements between 180° and 270°.

Angles in quadrant 4 have measurements between 270° and 360°.

What is the reference angle for an angle that measures 255°?

First, determine which quadrant in which the angles falls.

If an angle measures 255°, it falls within quadrant 3.

Next, use the formula: x-180, with x=255

255-180=75

The reference angle for 255° will be 75°.

Let's practice!

1. What is the reference angle for an angle that measures 56°?
2. What is the reference angle for an angle that measures 112°?
3. What is the reference angle for an angle that measures 308°?
4. What is the reference angle for an angle that measures 71°?
5. What is the reference angle for an angle that measures 194°?

Answers:

1. 56°
2. 68°
3. 52°
4. 71°
5. 14°

Signs of Trigonometry Values by Quadrant

The sign of a trigonometric value is determined by its point of the terminal side of the angle. Because some planes of the coordinate plane have negative locations, the calculated trigonometric function could be positive or negative.

This is easiest to calculate in Quadrant 1, where both the x and y values are positive. Because the x and y are both positive, all trigonometric functions are positive.

In Quadrant II, where the x value is negative and the y value is positive. The positive y value ensures a positive sine and cosecant. The negative x value ensures a negative cosine, tangent, secant and cotangent.

In Quadrant III, the x and y values are both negative. These negative values cause the sine, cosine, cosecant and secant to be negative. Also, these negative values cancel and lead to a positive tangent and cotangent.

In Quadrant IV, the x value is positive and the y value is negative. This will lead the sine, tangent, cosecant, and cotangent to be negative while the cosine and secant will be positive.

When you are given the signs of values, you can use that information to determine the quadrant where the point occurs and then apply that information to determine how large the angle is.

The cosine value of angle is negative and the sine value is positive. Its tangent is negative as well. Which choice could be a possible measurement for the angle?

a. 46°
b. 195°
c. 110°
d. 321°

Angles with a negative cosine and tangent and a positive sine are located in quadrant II. Angles in quadrant II fall between 90° and 180°. The choice that falls between 90° and 180° is choice C, 110°.

Let's practice!

1. The cosine value of angle is positive and the sine value is positive. Which choice could be a possible measurement for the angle?

a. 164°
b. 212°
c. 305°
d. 62°

2. The tangent value of angle is negative and the secant value is positive. Which choice could be a possible measurement for the angle?

a. 348°
b. 13°
c. 117°
d. 212°

3. The secant and cotangent value of angle is negative. Which choice could be a possible measurement for the angle?

a. 28°
b. 98°
c. 181°
d. 280°

4. The tangent value of angle is negative and the sine value is positive. Which choice could be a possible measurement for the angle?

a. 83°
b. 108°
c. 316°
d. 292°

5. The tangent value of angle is positive and the sine value is negative. Which choice could be a possible measurement for the angle?

a. 97°
b. 35°
c. 291°
d. 208°

Answers:

1. D
2. A
3. B
4. C
5. D

Graphing the Trigonometric Functions

The sine function of a graph looks like this:

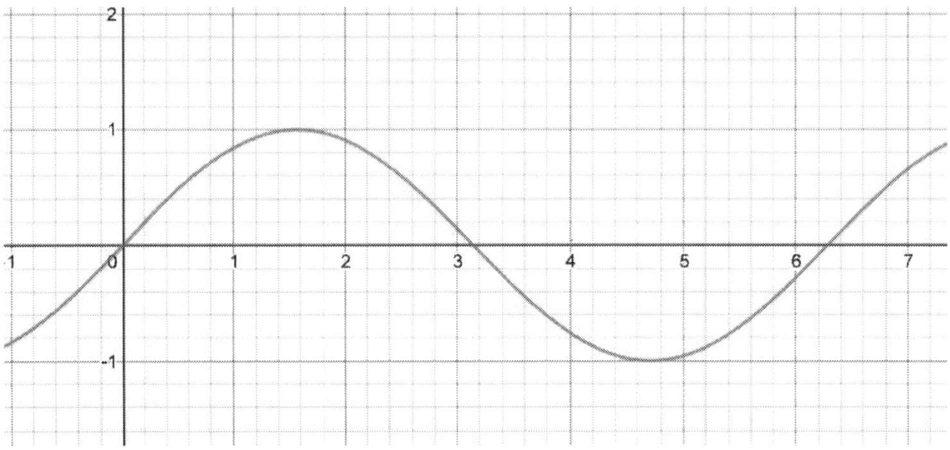

There are a few things to keep in mind about the sine graph.

The first is its shape. It has a curved shape that repeats.

Also, notice the graph intersects the y-axis at the origin. In addition, the graph intersects the x-axis at (0,0) and every multiple of after.

The cosine function of a graph looks like this:

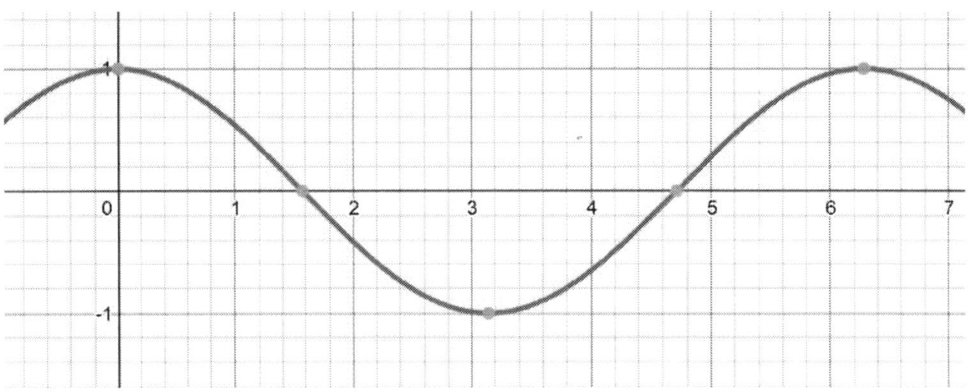

The cosine function is identical to the sine function except it is shifted left by 90 degrees or 2π radians.

The tangent function of a graph looks like this:

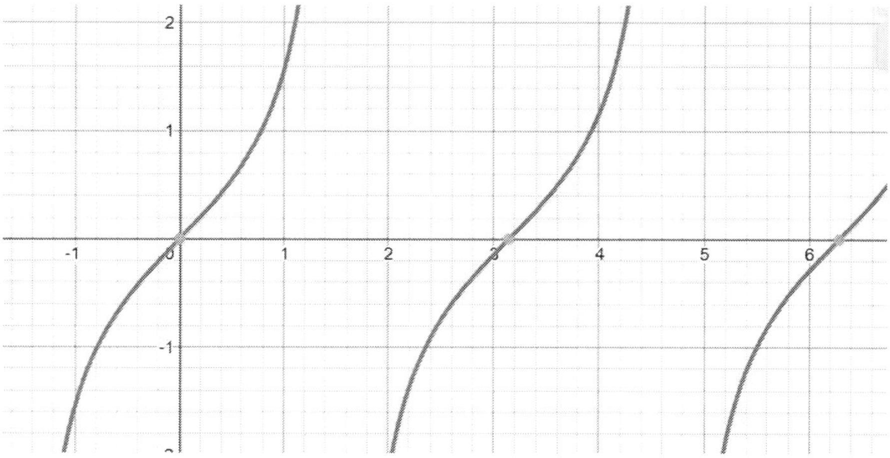

Because tangent is calculated by dividing the sine of x by the cosine of x, there are some values where the tangent function does not exist at. This is because there are some calculated values of sine x divided by cosine x that are undefined.

Also, the tangent function intersects the y axis at only one place (0,0) but it intersects the x axis at every positive and negative multiple of pi.

Finally, the graph of the tangent repeats at every radians, which is double the repetition of the graph of the sine and cosine function.

Let's review!

1. Which graph repeats at every radians?
2. Which graph intersects the x-axis at (0,) and exists as a function at approximately (0.9, 2)?
3. Which graph exists at (-1,)?
4. Which graph is not continuous as it has values that are undefined?
5. Which graph intersects the origin and has y-values that do not get larger than 1 or smaller than -1?

Answers:

1. Tangent
2. Sine
3. Cosine
4. Tangent
5. Sine

Properties of Graphs of Trigonometric Functions

Before we review the properties of specific functions, let review some graphic terms.

A period is the value of times a function takes to cover its complete value. Another way it is also considered as the amount of time needed before a function begins to repeat. Look at the graph below.

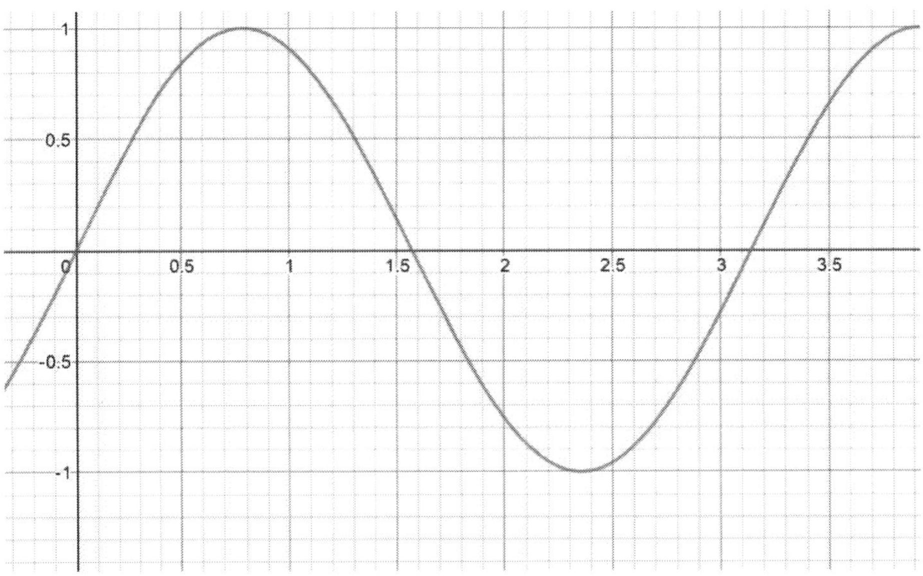

This function has new y values for the x values of 0through . But after , the calculated y values start to repeat again. Therefore, the period value of this function can be defined as .

The domain of a function is the set of all x values for the function which are defined. A function can contain a domain of all numbers, or it can contain a domain with specific values.

Consider the graph below.

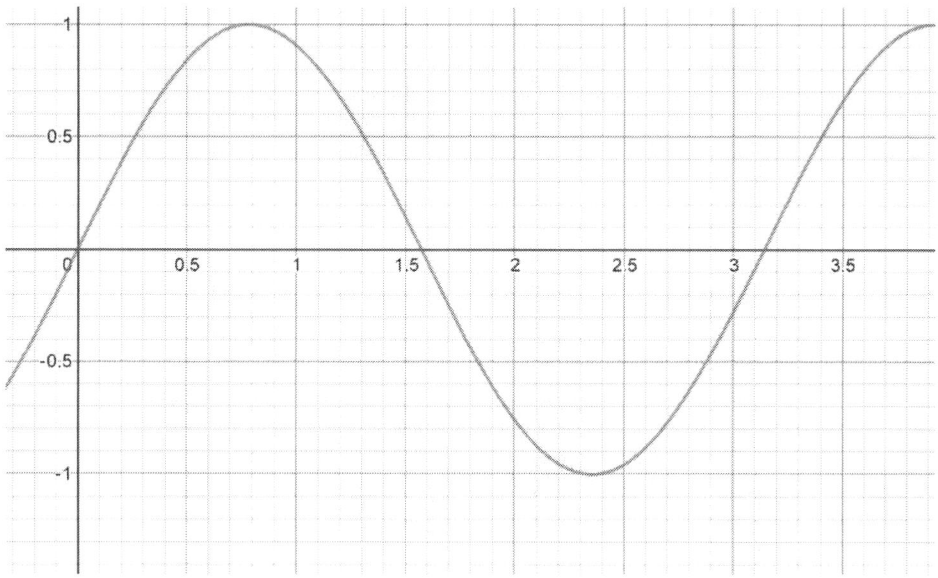

Values exist for this function for all negative and positive values of x. Therefore, the domain of this function is all real numbers.

The range of a function is all y-values of a function that can be calculated using values of x. A function can contain a range of all real number values or it can contain a specific set of values that are calculated using x.

Consider the graph below.

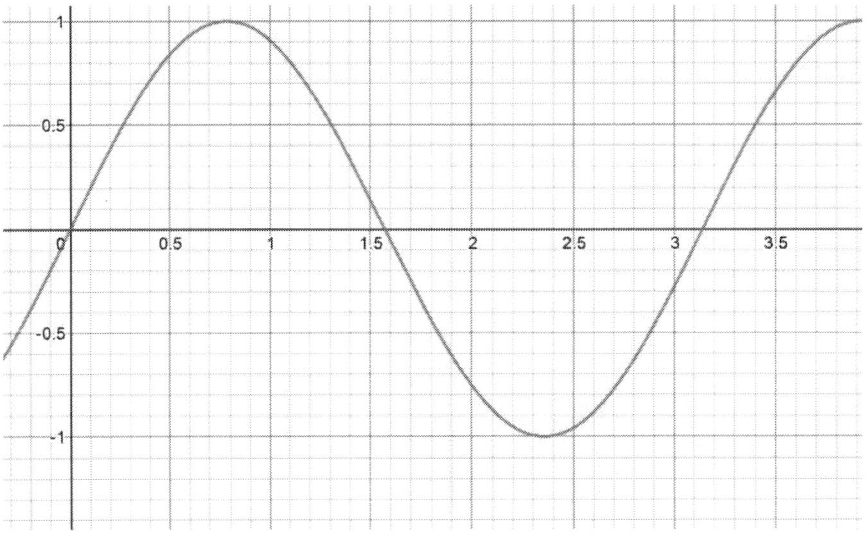

While its x values consist of all real numbers, it's y values are constrained between -1 and 1. Therefore, the range of this function is from -1 to 1.

Some functions have an amplitude. The amplitude of a periodic curve can be considered its vertical distance from the x-axis. It is calculated as the highest and lowest y values divided by 2.

Consider the graph below.

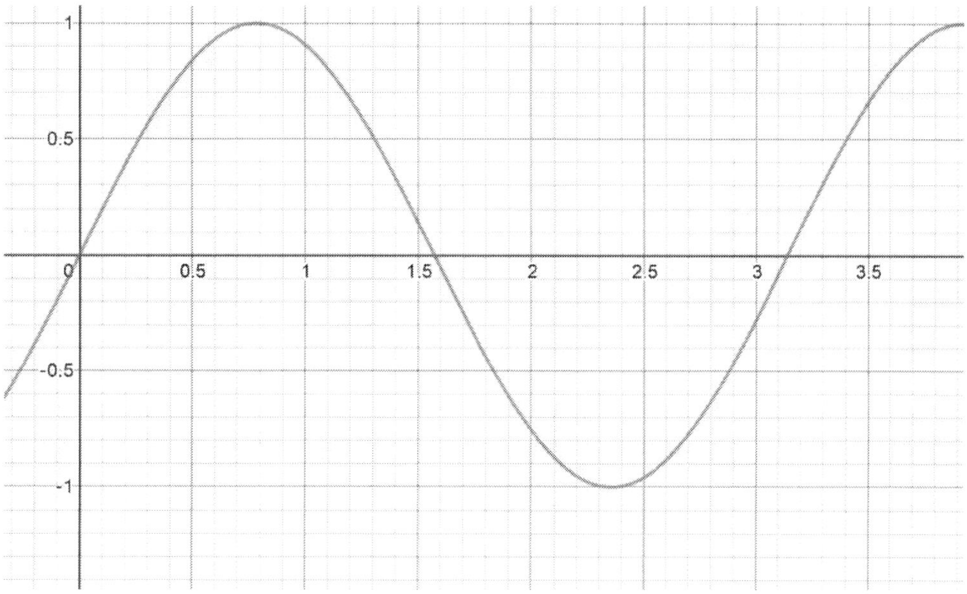

The highest y value is 1 and the highest low value is -1. The amplitude of this function is 1. (-1-1)/2 =-1 (taking the absolute value of this value gives us an amplitude of 1.)

Let's practice!

1. Find the period of the function below.

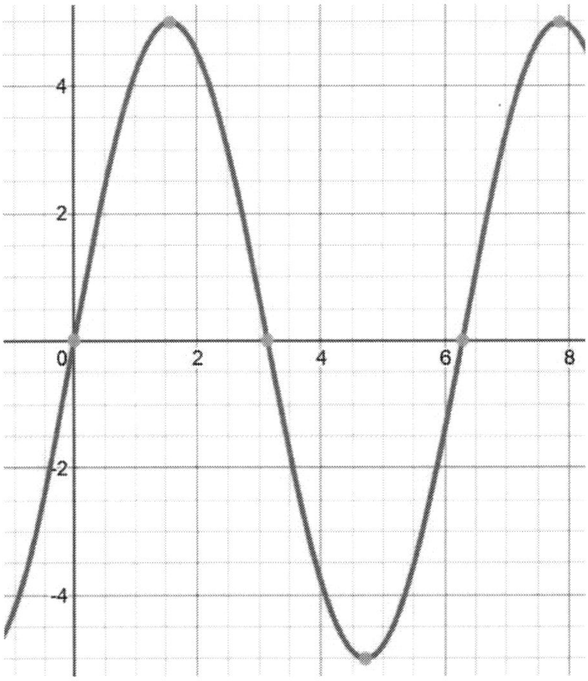

2. Find the range of the function below.

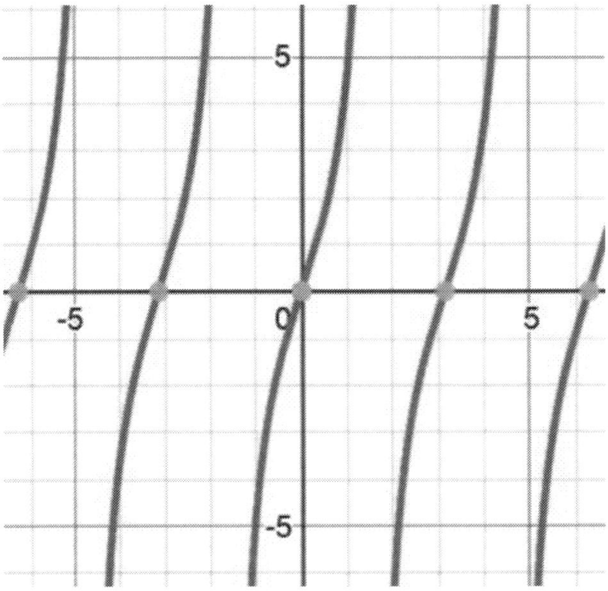

3. Find the amplitude of the function below.

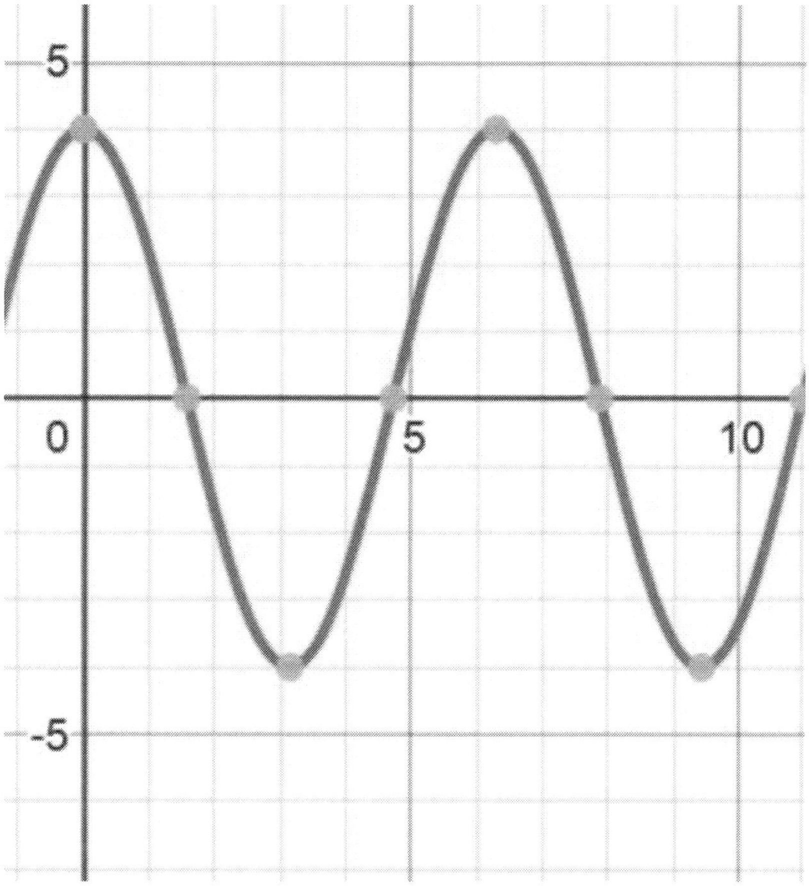

4. Find the domain of the function below.

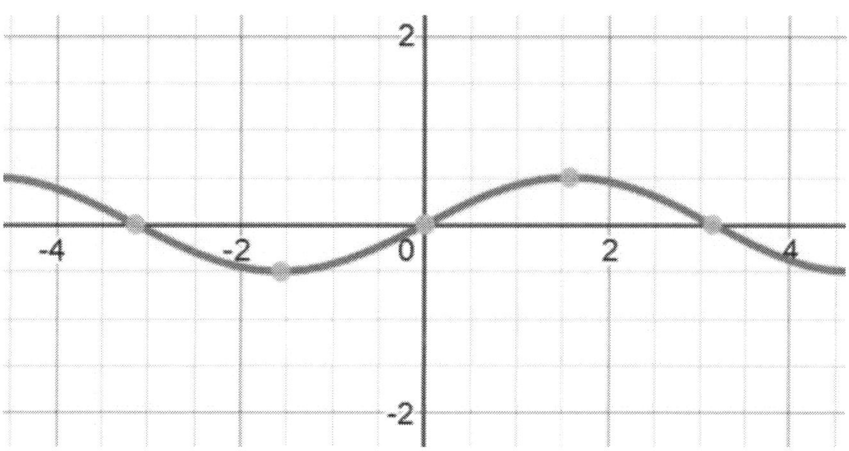

5. Find the range of the function below.

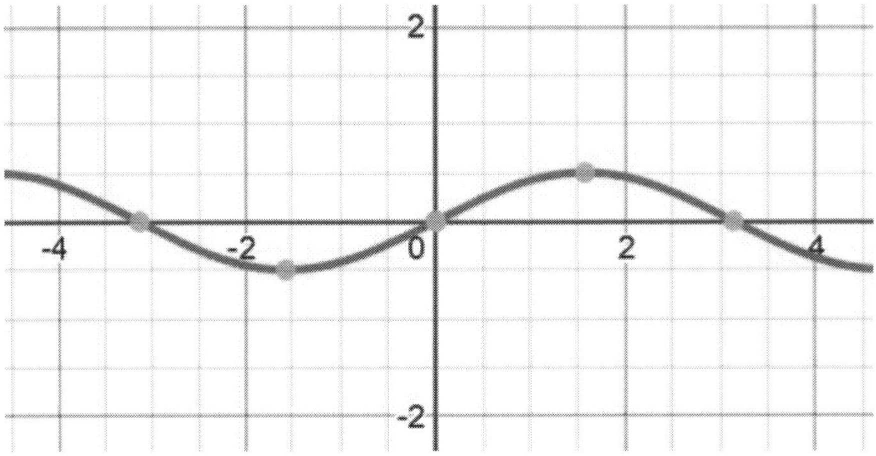

Answers

1. 2
2. All real numbers
3. 4
4. All real numbers
5. (-1/2, 1/2)

Finding an Equation from Its Graph

Any graph can be analyzed to get an equation from the information. For sine and cosine graphs, the formula is the following: y = a sin(bx + c) OR y = a cos(bx + c).

The value of a is defined as the amplitude. The value of b is defined as the period.

The value of c is the phase shift. Remember, the phase shift is how far left or right the graph has shifted. The phase shift should be calculated by comparing the graph to an original cosine or sine graph.

Let's review with an example.

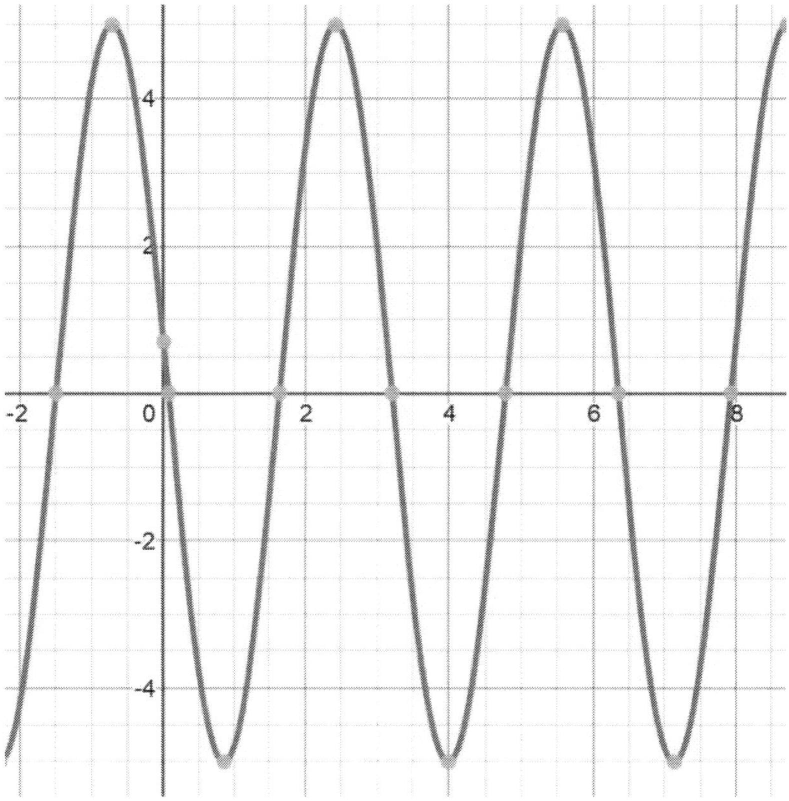

y = a sin(bx + c)

a= 5

b=2

c=3

Therefore, the graph can be written as

y=5 sin(2x+3)

Let's practice!

1. Find the graph of the image below.

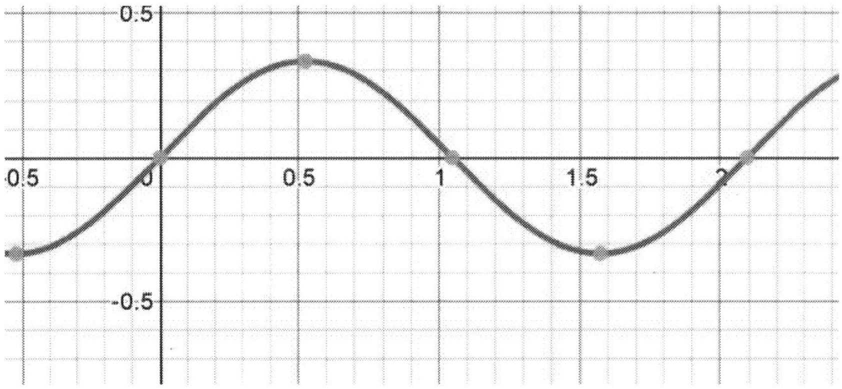

2. Find the graph of the image below.

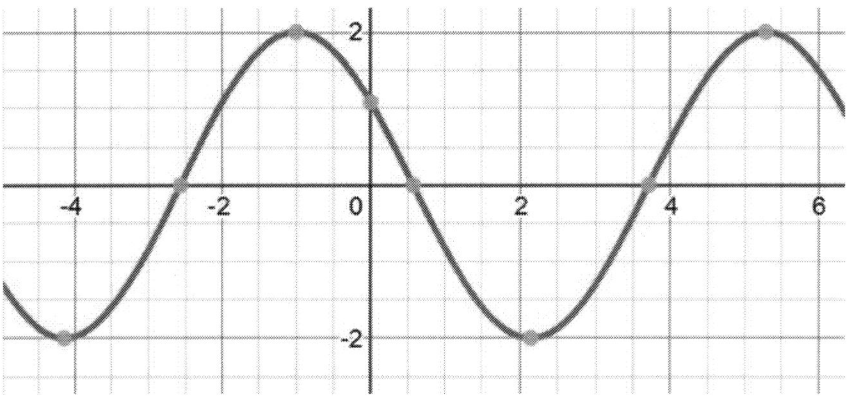

3. Find the graph of the image below.

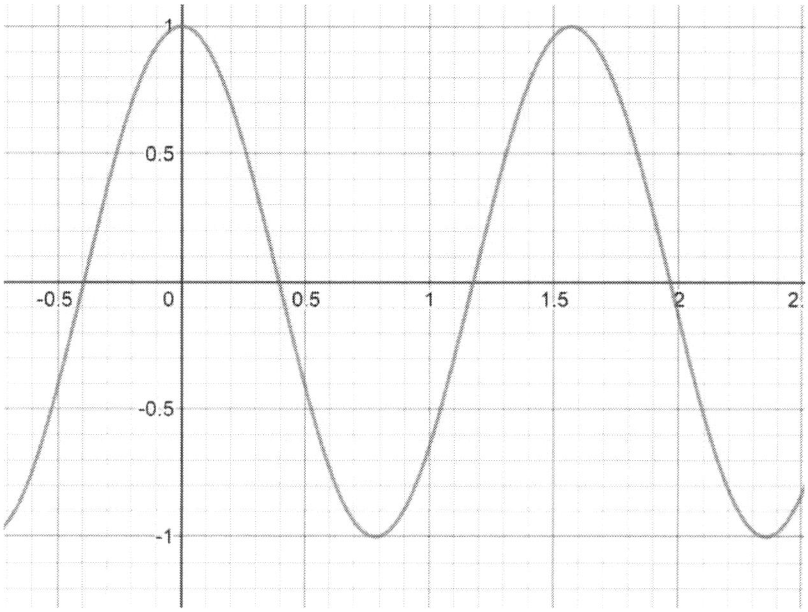

4. Find the graph of the image below.

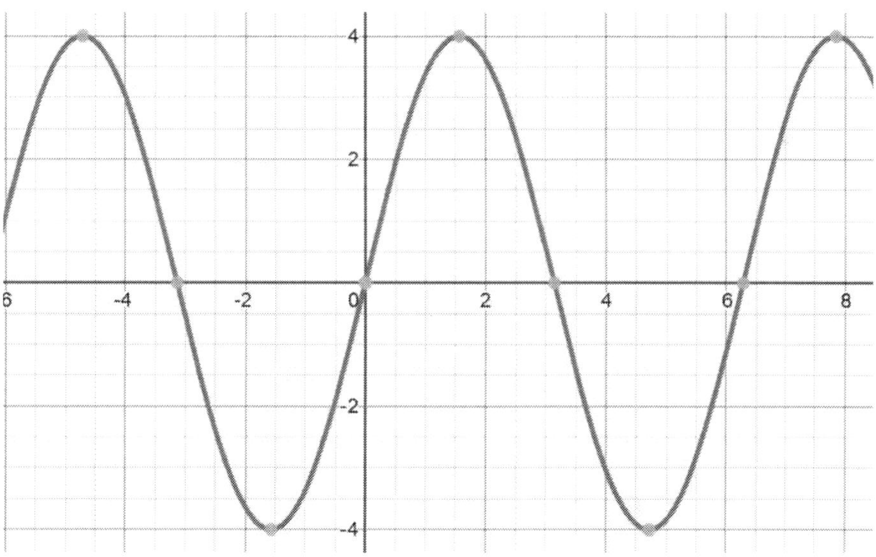

5. Find the graph of the image below.

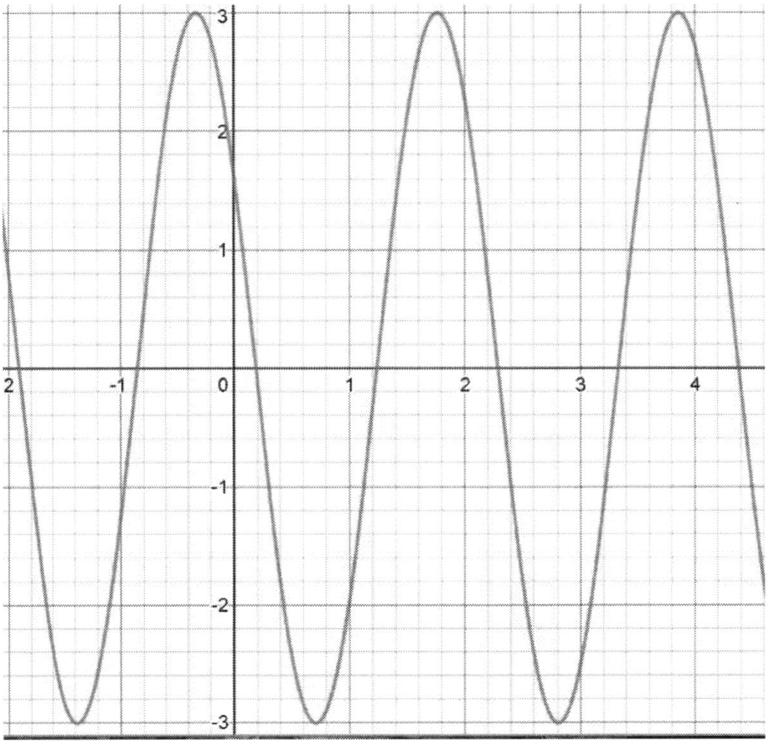

Answers:

1. y=⅓ sin 3x
2. y=2 cos (x+1)
3. y=cos 4x
4. y=4sin x
5. y=3cos (3x+1)

Inverse Trigonometric Functions

Remember the cosecant is defined as the inverse sine function or 1/sin x.

This means, for the cosecant function, the values are all values that are not included in the sine function.

Review the graphs below.

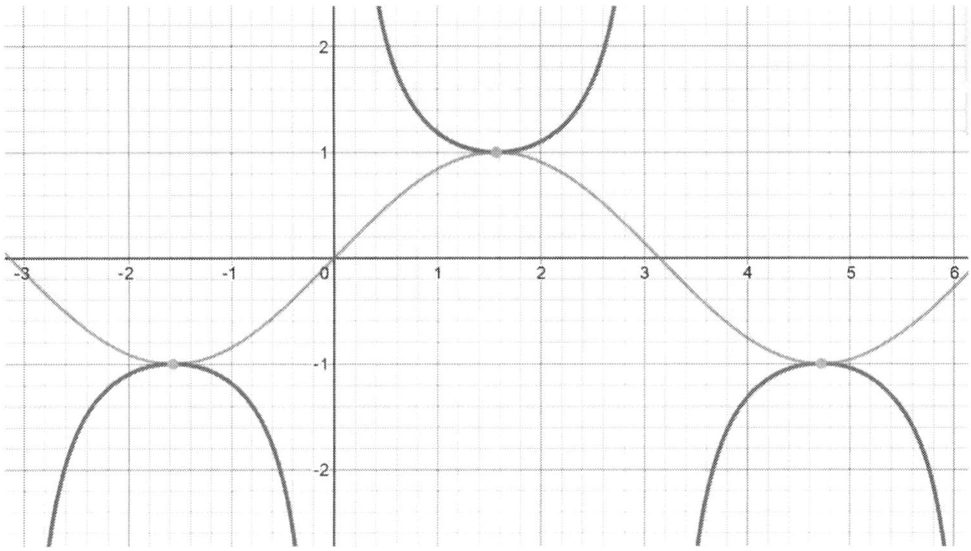

The red graph is the sine function.

The blue graph is the cosecant function.

You should notice that values exist in the cosecant function when they do not exist in the sine function.

The secant is defined as the inverse of the cosine function or 1/cosine x.

This means, for the secant function, the values are all values that are not included in the cosine function.

Review the graphs below.

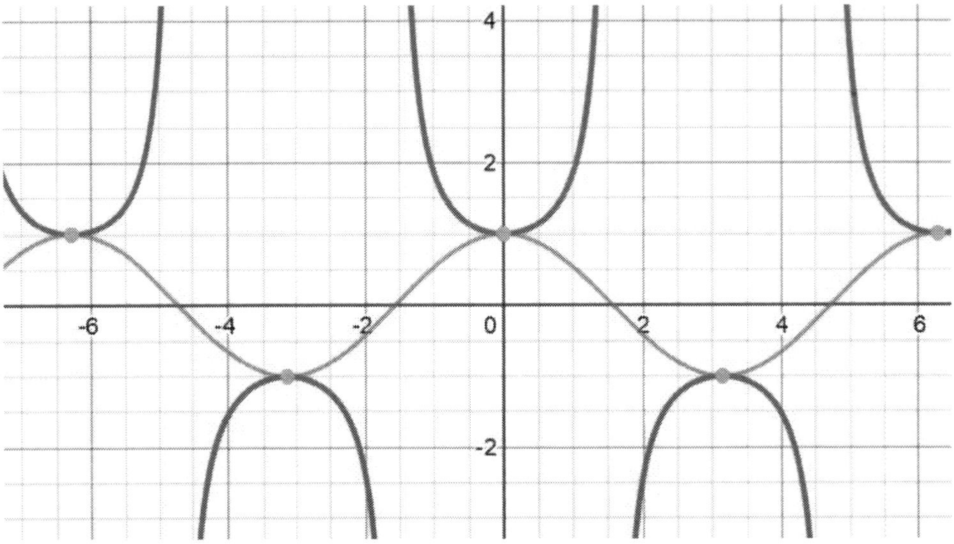

The red graph is the cosine function.

The blue graph is the secant function.

You should notice that values exist in the secant function when they do not exist in the sine function.

Remember the cotangent is defined as the inverse tangent function or 1/tan x.

This means, for the cotangent function, the values are all values that are not included in the sine function.

Review the graphs below.

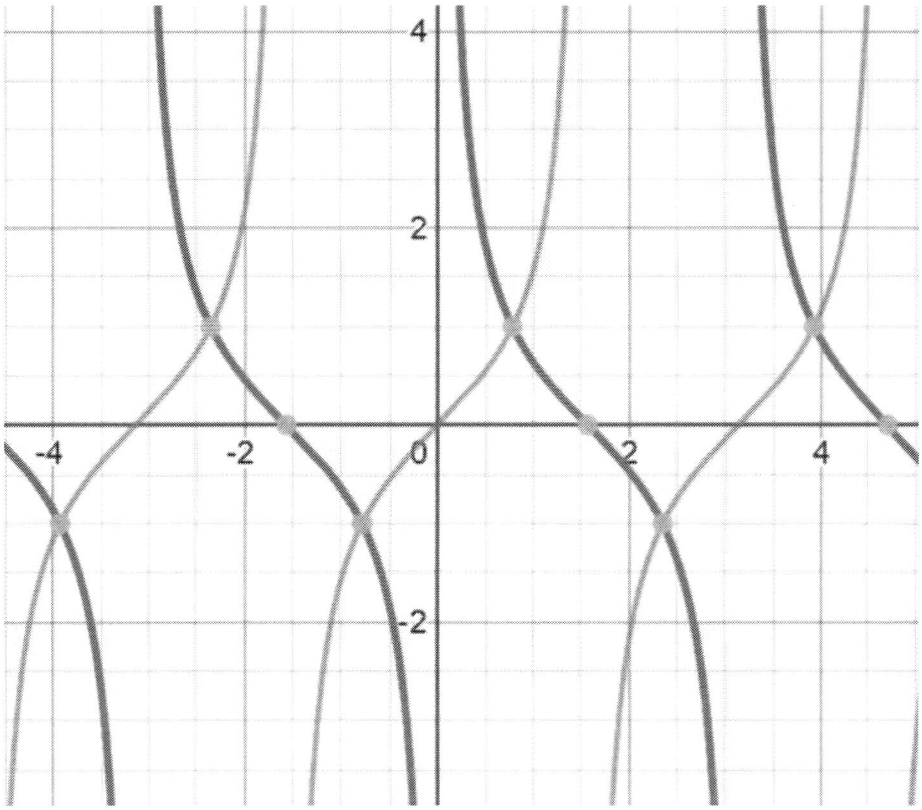

The red graph is the tangent function.

The blue graph is the cotangent function.

You should notice that the values of the tangent and cotangent functions are opposites.

Let's practice.

Identify what type of graph is identified.

1. 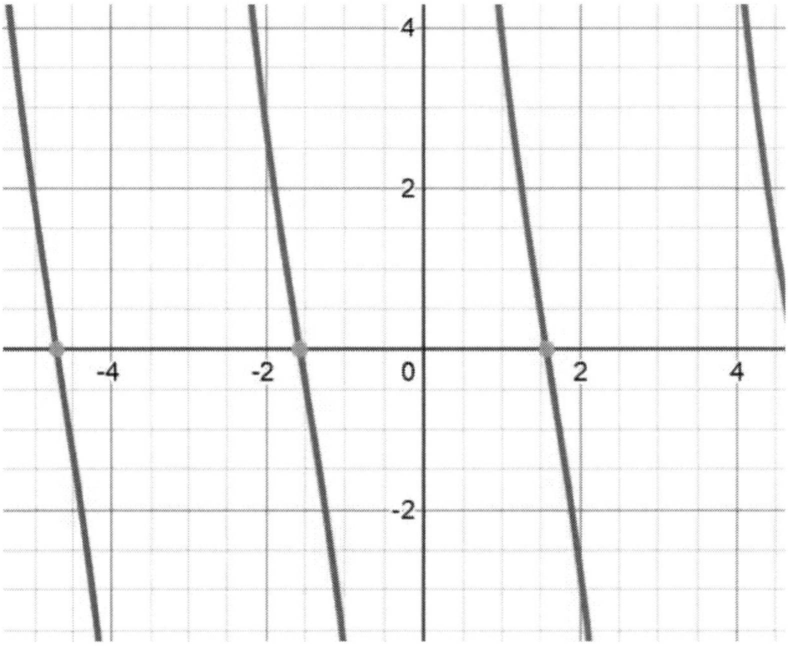

2. 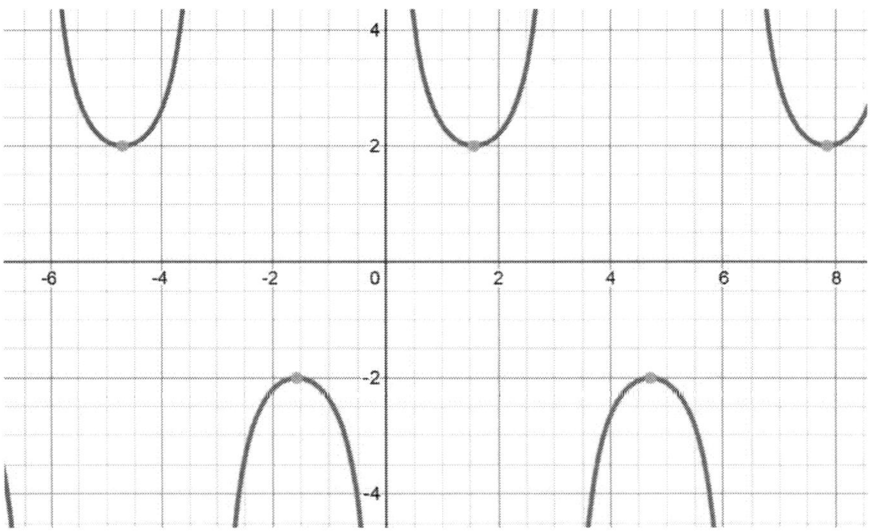

3. 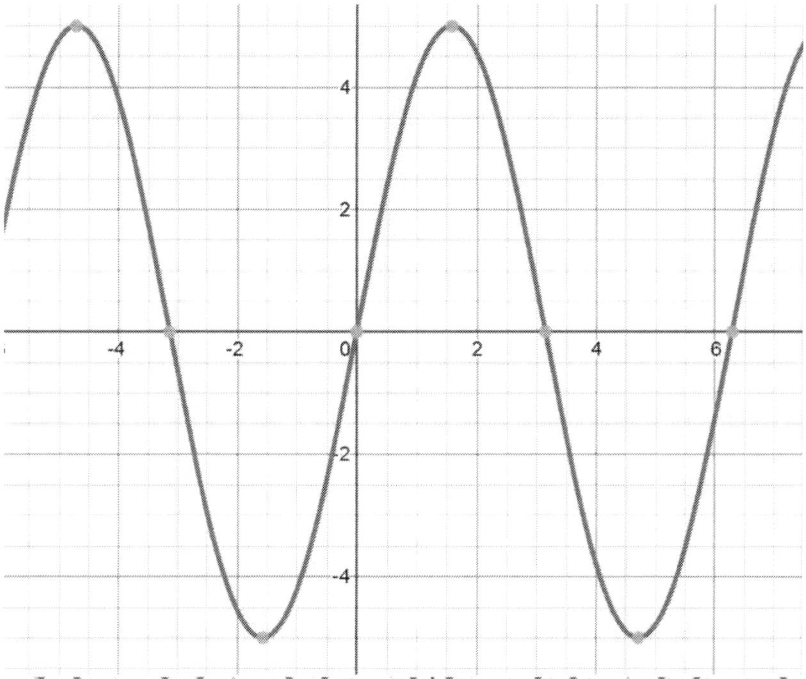

4. 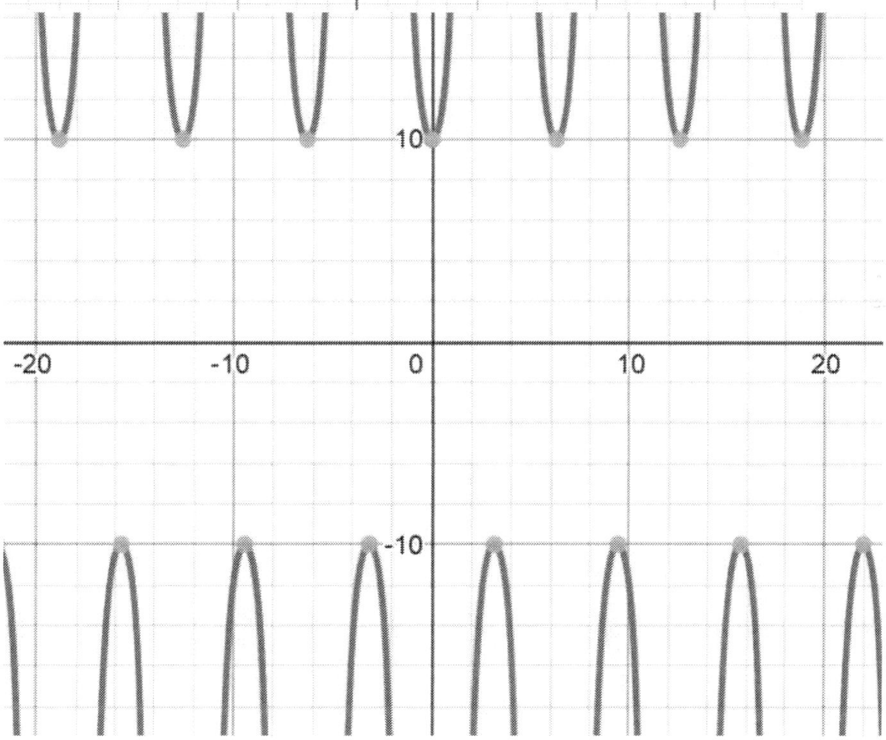

5. 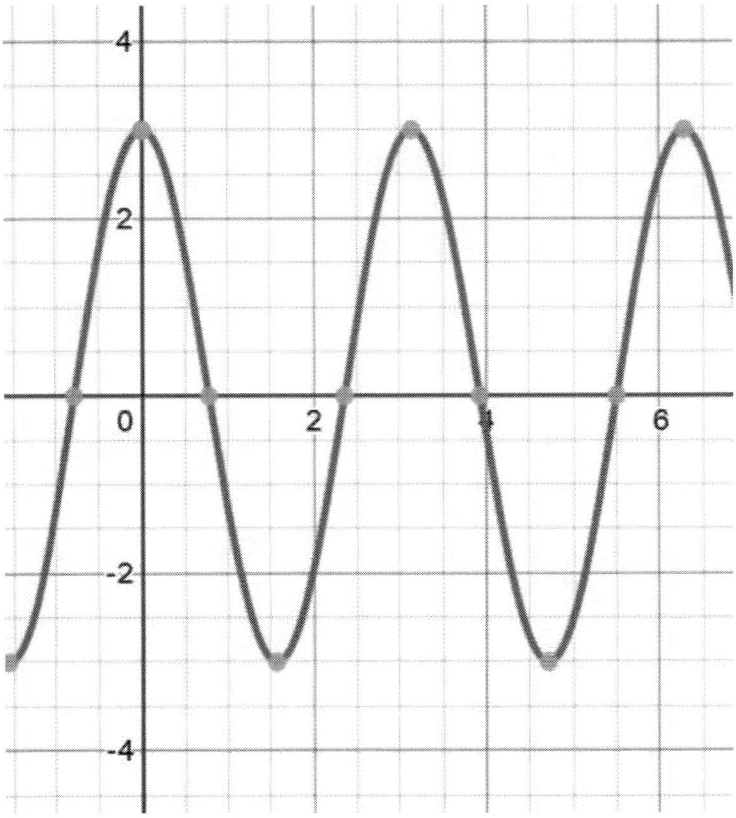

Answers

1. Cotangent
2. Cosecant
3. Sine
4. Secant
5. Cosine

Basic Trigonometric Identities

Now that you have a good understanding of the basic identities, let's take a look at how they can be applied to find other information related to the Pythagorean Theorem. Let's start first with the related identities.

sin2 (A) + cos2 A= 1

tan2(A) +1= sec2A

1+cot2(A)=csc2A

If you are given information on some of these identities, you can use that information to find other identities.

As you solve the problems, remember that sin2A, is found by calculating the sin A and then squaring that answer. Also, sin 2A, is found by calculating the sin A and then doubling your result.

Let's look at an example:

Prove the first theorem using the angle of 41°.

sin 2 A + cos2 A= 1

(sin 41)2 + (cos 41)2= 1

(0.65605903)2 + (0.75470958)2= 1

0.43041345+0.569586=1

1.00000…=1

Problems

1. Prove 1+cot2A=csc2A using an angle of 18° showing every step.
2. Prove tan2A +1= sec2A using an angle of 68° showing every step.
3. Prove sin2 A + cos2 A= 1 using an angle of 43° showing every step.

4. Prove tan2A +1= sec2A using an angle of 51° showing every step.
5. Prove 1+cot2A=csc2A using an angle of 26° showing every step.

Answers:

Each problem should show each step of the proof with the calculated value being very close to 1.

The Major Formulas

How to Solve

There are also identities that help define how angles relate to each other. For sine, cosine and tangent, the sine/cosine/tangent of a negative angle is the same value as the negative of the sine/cosine/tangent of the same positive angle.

Or

sin -=-sin .

There are also identities that relate the double values of an angle to other calculated values.

These are the double-angle identities or formulas.

sin 2 =2sin x 2cos

It also can be calculated as 2 tan 1+tan2.

cos 2 =cos2 -sin2

It also equals 2cos2 -1, 1-2sin2 and 1-tan2 1+tan2

tan 2=2 tan 1-tan2

Let's review by matching the identities.

1. 1-tan2 1+tan2	A. sin 2
2. 2 tan 1-tan2	B. -cos

3. cos -   C. tan 2

4. 2sin x 2cos   D. 2cos2 -1

5. 1-2sin2   E. cos2 -sin2

Answers

1. D, E
2. C, A
3. B
4. A, C
5. D, E

Additional Identities

How to Solve

The Law of Sines states that asin A=bsin B=csin C

This means that any side divided by the sine of the angle of that side is equal to any other side divided by the sine of its angle.

For half of a provided angle, you can use the half-angle formulas.

sin 2=1- cos 2

cos 2=1+ cos 2

tan 2=1- cos 1+ cos =sin 1+ cos

There are also the addition formulas.

sin (A B)=sin A cos B cos A sin B

cos (AB) = cos A cos B sin A sin B

Practice!

Let's review the formulas! Match the calculations which are equal.

1. sin A cos B cos A sin B   A. bsin B
2. 1+ cos 2   B. cos A cos B sin A sin B

3. csin C
4. cos (AB)
5. sin 1+ cos

C. sin (A B)
D. 1- cos 1+ cos
E. cos 2

Answers

1. C
2. E
3. A
4. B
5. D

Simple Trigonometric Equations

How to Solve

A trigonometric equation is any equation that has a variable inside a trigonometric function.

Sin x=4 is a trigonometric equation, Sin 4= x is not.

You can solve trigonometric equations using your knowledge of trigonometry and Algebra.

Remember, to solve linear equations, you isolate the variable on one side of the equation to find its value.

5x+2=12

5x+2-2=12-2

5x=10

5x5=105

x=2

You can do this with trigonometric functions also!

Let's look at an example.

-4+cos x =-3

To solve, we should add 4 to each side.

Cos x=1

When the cosine x=1, we know the value of the angle is 0.

You can also review your knowledge of the quadratic formula to solve trigonometric quadratic equations.

Remember, any quadratic equation can be solved using the values of the quadratic formula.

The quadratic formula is x=-bb2-4ac2a

When using the quadratic formula, the trig function replaces the variable part of the equation.

For example, $\sin^2 x + 7\sin x - 6 = 0$

To use the quadratic formula, you identify the a, b and c values.

In this example, a=1, b=7 and c=-6

Now you work to solve.

sin x=-bb2-4ac2a

sin x=-772-4(1)(-6)2(1)

sin x=-749+242

sin x=-7642

sin x=-782

sin x=-7+82sin x=-7-82

sin x=12sin x=-152

Sin x is 0.5 OR -7.5

The second value does not make logical sense but sin x=0.5 works.

If sin x=0.5, then the x value is around 0.785.

If you cannot use the quadratic equation, you can use your knowledge of Algebra and trigonometry to solve for an unknown.

Let's look at an example:

8 tan2x=4

As you do with a variable, you want to get the squared part on one side so divide both sides by 8.

tan2x=12

Using the inverse tangent, you get x=0.464

Trigonometric Equations with Multiple Angles

Remember, trigonometric functions are periodic, meaning they repeat.

Because of this when there are multipliers, the number of solutions increases.

A multiplier is shown as a coefficient to the variable taken to the trigonometric function, such as tan 5x.

When you solve, you can take these equations apart and see how the solutions repeat.

Let's look at an example:

Sin 6x=0

6x=sin-1(0)

6x6=0, , 2, 3, 4, 5, 66

x=0, 6, 3,2, 56,

Let's Practice!

1. 7+cos x=6
2. 14= 14=sin x
3. $\cos^2 x$ -2cos x -8 = 0
4. $\sin^2 x$ +sin x – 10= 0
5. sin2x4-6=0

Answers

1. x=180
2. x=0
3. x=¼
4. x=⅙
5. No solution

Oblique Triangle Laws

When we have any size triangle, you can use the Law of Sines and the Law of Cosines to solve triangles.

The Law of Sines states:

$$\frac{a}{\sin A} = \frac{b}{\sin B} = \frac{c}{\sin C}$$

A, b and c are sides of a triangle and the angles of which you are taking the sine are the sides opposite each angle.

In words, the Law of Sines says, we can take the side length of side a and divide it by the sine of angle A that amount equals the side length of side b divided by the sine of angle B and the side length of side c divided t by the sine of angle C.

Let's see how this works.
Review the information given for the triangle below.

A triangle has a measurement of 56° for angle A and a measurement of 63° for angle B. We are not given the measurement for angle C. We are given a side length of 9 for side a and they want us to find side b.

What should we do?

Let's review our formula.

$$\frac{a}{\sin A} = \frac{b}{\sin B} = \frac{c}{\sin C}$$

What do we know? Angle A, side a and Angle B.

$$\frac{a}{\sin A} = \frac{b}{\sin B}$$

Let's fill in the values we know.

$$\frac{9}{\sin 56} = \frac{b}{\sin 63}$$

Let's use our equation solving skills to rearrange:

$$\frac{9\,(\sin 63)}{\sin 56} = \frac{b\,\cancel{(\sin 63)}}{\cancel{\sin 63}} = \frac{c}{\sin c}$$

The "sin 63" cancel on the right hand side, so now we just need to use our calculator to solve!

The missing side is 9.7 units!

We can also use the Law of Sines to find an unknown angle! Let's look at a triangle- we are given an angle measurement for angle C of 65°. We are also given a measurement for side c of 5.9 units and a measurement for side b of 4.2 units. Let's use the Law of Sines to calculate angle B.

What should we do?

Let's review our formula.

$$\frac{a}{\sin A} = \frac{b}{\sin B} = \frac{c}{\sin C}$$

When we are looking for an unknown angle, it is best to turn the Law of Sines upside down.

$$\frac{a}{\sin A} = \frac{b}{\sin B} = \frac{c}{\sin C}$$

becomes

$$\frac{\sin A}{a} = \frac{\sin B}{b} = \frac{\sin C}{c}$$

What do we know? Angle C, side c and side b.

$$\frac{\sin B}{b} = \frac{\sin C}{c}$$

Let's fill in the values we know. sin B4.2= sin 655.9

$$\frac{\sin B}{4.2} = \frac{\sin 65}{5.9}$$

Let's use our equation solving skills to solve:

$$\frac{\cancel{4.2}(\sin B)}{\cancel{4.2}} = \frac{\sin 65(4.2)}{5.9}$$

Multiplying each side by 4.2 removes that value from the left side.

Sin B= 0.6451...

Next, we use the inverse sine function to find the angle.

Angle B is around 40.1°.

There is one thing we need to consider. Sometimes, when we have 1 angle and 2 sides, there are actually two possible answers. We can have a scalene triangle with three small angles but we can also have an obtuse triangle with 2 small and 1 large angle.

This only happens when you are given two angles with a side not between the two angles, although not always.

When you are solving the problem, you have to consider, could I swing the side the other way and also make a correct answer?

Let's look at an example.

We are given the information for a triangle: measurements are provided for angle A and side a. Angle A measures 34° and side a measures 17. The side beside angle A and side a is side b. You are also given the measurement for side b, 26 units.

Let's review our formula.

$$\frac{a}{\sin A} = \frac{b}{\sin B} = \frac{c}{\sin C}$$

When we are looking for an unknown angle, it is best to turn the Law of Sines upside down.

$$\frac{\sin A}{a} = \frac{\sin B}{b} = \frac{\sin C}{c}$$

What do we know? Angle A, side a and side b. $\frac{\sin A}{a} = \frac{\sin B}{b}$

Let's fill in the values you know. $\frac{\sin 34}{17} = \frac{\sin B}{26}$

Let's use our equation solving skills to solve:

$$\frac{26(\sin 34)}{17} = \frac{\cancel{26} \sin B}{\cancel{26}}$$

Multiplying each side by 26 removes that value from the left side.

Sin B= 0.855346...

Next, we use the inverse sine function to find the angle.

Angle B is around 58.8°.

There is another angle that gives you a sin of 0.8555. To find it, we take the found angle away from 180.

Angle B could also be 121.2°.

Let's use the Law of Sines and practice. Use the given measurements to try and find the missing value for each triangle.

Let's practice!

1. Angle A measures 52° and angle B measures 22°. Side b measures 7. What does side a measure?
2. Angle A measures 64° and angle B measures 37°. Side c measures 6.2. What does side b measure?
3. Angle A measures 49° and angle B measures 21°. Side b measures 3.8. What does side c measure?
4. Angle B measures 72°. Side b measures 11 and side c measures 4. What does Angle C measure?
5. Angle B measures 113° and side b measures 14.2. Side c measures 5.2. What does Angle C measure?
6. Angle A measures 34° and side a measures 7.9. Side b measures 11.9. Angle A is NOT between sides a and b. What are the two possible measurements for angle B?

Answers (rounded to the nearest tenth)

1. 14.7 units
2. 3.8 units
3. 10 units
4. 20.2°
5. 19.7°
6. 70.2° and 109.8°

Let's look at the Law of Cosines.

The Law of Cosines states $c^2 = a^2 + b^2 - 2ab\cos(C)$.

When should we use the Law of Cosines? If we are given 2 sides and the angle between them and we want to find the third side, the Law of Cosines will work.

You are given the measurements for side a (7 units) and side b (12 units) and the measurement for angle C, 49°. What is the measurement for side c?

Let's use the Law of Cosines to find side c!

$c^2 = 7^2 + 12^2 - 2(7)(12)\cos(49)$.

Simplify using order of operations: $c^2 = 49 + 144 - 110.217$

$$c^2 = 82.783$$

To find c, you should take the square root of both sides:

$$\sqrt{c^2} = \sqrt{82.783}$$

c= 9.1 units

The law of cosines is also useful to use when we are given three sides of a triangle.

A triangle has 3 measurements: side a measures 11, side b measures 9 and side c measures 7. What is the measurement of Angle C?

Let's use the Law of Cosines!

$$c^2 = a^2 + b^2 - 2ab\cos(C)$$
$$7^2 = 11^2 + 9^2 - 2(11)(9)\cos(C)$$

Let's use order of operations and algebra to solve.

49=121+81-198 cos C

49-121-81=-198cos C

-153= -198cos C

0.7727 = cos C

Use the inverse cosine function to find the measurement of angle C to be 39.4°.

We can rewrite the Law of Cosines formula to help you find the measurements of angles A and B also:

$$a^2 = b^2 + c^2 - 2bc\cos(A)$$
$$b^2 = a^2 + c^2 - 2ac\cos(B)$$

Let's use the Law of Sines and practice. Use the given measurements to try and find the missing value for each triangle.

Let's practice!

1. Side a of a triangle measures 3 units and side c measures 9 units. Angle B, which is in between sides a and c, measures 72°. What is the measurement of side b?
2. Side a of a triangle measures 2.8 units and side b measures 4.9 units. Angle C, which is in between sides a and b, measures 56°. What is the measurement of side c?
3. Side a of a triangle measures 8 units and side c measures 19 units. Angle B, which is in between sides a and c measures 110°. What is the measurement of side b?
4. You are given 3 sides of a triangle, side A is 8, side B is 4 and side C is 11. What is the measurement of angle A?
5. You are given 3 sides of a triangle, side A is 5.3, side B is 9.6 and side C is 7.1. What is the measurement of angle B?
6. You are given 3 sides of a triangle, side A is 4, side B is 10 and side C is 8. What is the measurement of angle C?

Answers (rounded to the nearest tenth)

1. 8.6
2. 4.1
3. 23
4. 33.9°
5. 100.5°
6. 49.4°

A complex number is made up of a real number and an imaginary number.

A real number is a number that is a whole number, a rational number or an irrational number.

Real numbers can have a positive or negative value as well as a zero value.

An imaginary number is any number that gives a negative result when squared.

We consider these imaginary because usually, when we square a positive or negative number, we get a positive result.

Imaginary numbers are based on the unit of i, which is the square root of -1.

When we square i, we get a result of -1.

Complex numbers are made when you combine a real number and an imaginary number.

Complex numbers look like:

A+bi, where a is a real number and bi is the imaginary number.

A complex number has a real part and an imaginary part. But in all numbers, either part can have a value of zero.

This means all numbers are complex numbers.

Real numbers are complex numbers with an imaginary part value of 0.

Imaginary numbers are complex numbers with a real part value of 0.

Imaginary numbers are represented on a number line that is vertical through the real number line at a location of 0.

You plot complex numbers similar to the way you plot coordinates on the coordinate plane. For example, given the number 7+2i, you use the horizontal real number line to move right 7 units, then use the vertical imaginary number line and go up 2 units. The point for 7+2i is located where these two locations intersect.

We can combine complex numbers by adding them. When we do so, we add each part separately. A good rule to follow is:

(a+bi) + (c+di) = (a+c)+(b+d)i

This is simply adding the real numbers and then adding the imaginary numbers.

Let's look at an example: What is (5+6i)+(4+9i)?

Add the real numbers: 5+4

And the imaginary numbers: 6+9

The answer is 9+15i.

We can multiply complex numbers too.

When we multiply, we use the FOIL method, which you may remember from multiplying binomials in Algebra: First, Outside, Inside, Last.

(a+bi) (c+di) becomes a x c + a x di + bi x c + bi x di

One note, when multiplying imaginary numbers, it is important to keep in mind the value of i^2 is -1.

Let's look at an example. What is (5+6i)(4+9i)?

(5)(4)   +(5)(9i)+(6i)(4)+ (6i)(9i)
20   + 45i   +24i   +54i2
Remember 54i2becomes 54(-1)
20+45i+24i-54
Combine like terms: -34+69i

Another way to represent a complex number is using its polar form. The form a+bi is the rectangular coordinate form of a number.

Looking at the graphic representation of a complex number, we can create a right angle, where the side a is the value of the real number component and the side b is the value of the imaginary number component. This leaves us with two unknowns we can use trigonometry to solve.

The third side, r, can be calculated using the Pythagorean Theorem: $r^2 = a^2 + b^2$

We can use trigonometric ratios to find the value of the angle that corresponds with the imaginary number side. Because of that, we know cos =ar and sin =br.

We can solve each function for their respective side: a and b. Once we do that, we know that $\cos\theta = \frac{a}{r}$ and $\sin\theta = \frac{b}{r}$.

We can use the coordinate form, z= a+bi, and substitute the values for a and b.

z=a+bi

$z = r(\cos\theta + i\sin\theta)$

Remember, you can find r with the Pythagorean Theorem: $r^2 = a^2 + b^2$ so r= $\sqrt{a^2 + b^2}$

Let's look at an example.
Express the complex number in polar form.   7+2i

Let's find the value of r first: $\sqrt{a^2 + b^2} = \sqrt{7^2 + 2^2} = \sqrt{53} \approx 7.28$

Since a>0, we can use the formula $\theta = \tan^{-1}\frac{b}{a}$ to find θ

$\theta = \tan^{-1}\frac{2}{7}$ =15.9

We can use the information we found to rewrite the complex number in polar form.

7.28(cos (15.9)+ i sin (15.9).

Let's practice!
How would you plot these imaginary numbers?

1. 2+4i
2. 1-9i
3. 3-7i
4. 8+5i
5. 6+12i

Let's combine these complex numbers.

1. (17+3i)+(3+2i)
2. (4-6i)+(5+7i)
3. (2+i)-(3-4i)
4. (8-9i)+(10+13i)
5. (5+2i)-(8-i)

Let's multiply these complex numbers.

1. (11-7i)(2+6i)
2. (3+8i)(4+i)
3. (6-5i)(8+9i)
4. (10-i)(12+4i)
5. (4+3i)(3+2i)

Rewrite these complex numbers in their polar form.

1. 4+5i
2. 2-6i
3. 1+3i
4. 7-2i
5. 8+i

Answers

How would you plot these imaginary numbers?

1. To the right of zero two, above the number line 4
2. To the right of zero one, below the number line 9
3. To the right of zero three, below the number line 7
4. To the right of zero eight, above the number line 5
5. To the right of zero six, above the number line 12

Let's combine these complex numbers.

1. 20+5i
2. 9+i
3. -1+5i
4. 18+4i
5. -3+3i

Let's multiply these complex numbers.

1. 64+52i
2. 4+35i
3. 93-14i
4. 124+28i
5. 6+17i

Rewrite these complex numbers in their polar form.

1. 6.43(cos (51.3)+ i sin (51.3)
2. 6.32(cos (-63.4)+ i sin (-63.4)
3. 3.16(cos (71.6)+ i sin (71.6)
4. 7.28(cos (-15.9)+ i sin (-15.9)
5. 8.06(cos (7.1)+ i sin (7.1)

Review Problems

Section 1: Introducing Trigonometry-The Distance Between Two Points

Calculate the distance between the two points listed.

1. (-4, 7) and (14, 6)
2. (11, -19) and (3, 8)
3. (14, -8) and (1, -8)
4. (3, -7) and (10, -2)
5. (14, 1) and (-13, 6)
6. (-8, -7) and (-4, 9)
7. (14, 1) and (-1, 14)
8. (-15,-20) and (42, 16)
9. (-4, -25) and (25, -4)
10. (14, -13) and (-11, 22)
11. (9, 22) and (-4, 14)
12. (2, -8) and (12, 10)

13. (21, -6) and (7, 32)

14. (10, -8) and (-1, 13)

15.(6, -11) and (3, -15)

Section 1: Introducing Trigonometry- Functions

Review the x and y values given for each relationship. Indicate whether the relationship is a function and if it is, determine the rule.

1. 2. 3.

Points: (7, 10); (11, 14) and (21, 24) Points: (-7, 5); (-10, 16) and (-7, 22)
Points: (13,-7); (-4, 3) and (12, 7)

4. 5. 6.

Points: (144, 12); (121, 11) and (48, 4) Points: (4, 8); (4, -6) and (4, 10) Points:
(2, 3); (4, 13) and (5, 18)

7. 8. 9.

Points: (-7, 6); (3, 4) and (-7, -6) Points: (12, 2); (10, 4) and (10, 6) Points:
(2, 8); (4, 64) and (8, 512)

10. 11. 12.

Points: (5, 11); (11, 5) and (5, -11) Points: (10, 3); (14, 7) and (20, 13) Points:
(12, 7); (2, -5) and (12, -11)

13. 14. 15.

Points: (3, 14); (6, 26) and (7, 30) Points: (2, 4); (5, 25) and (7, 49) Points:
(2, 9); (2, -5) and (12, -11)

Section 1: Introducing Trigonometry- Graphs of Functions

Review the graph and identify if the function is linear or quadratic.

1.

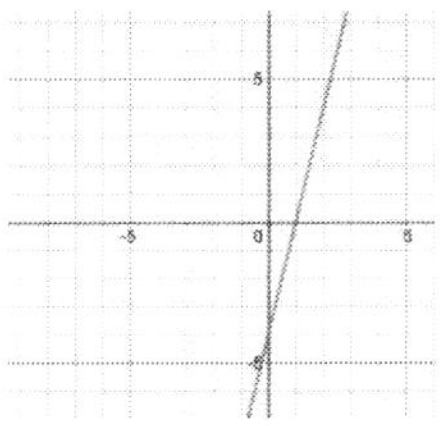

2.

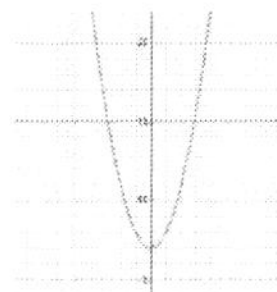

3.

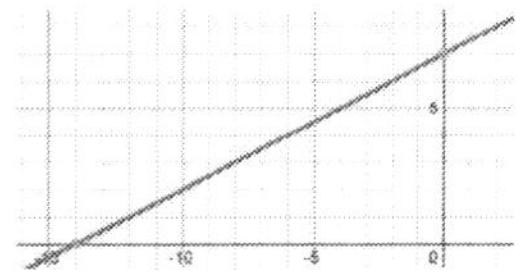

4.

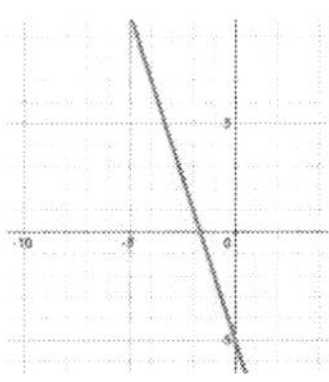

5.

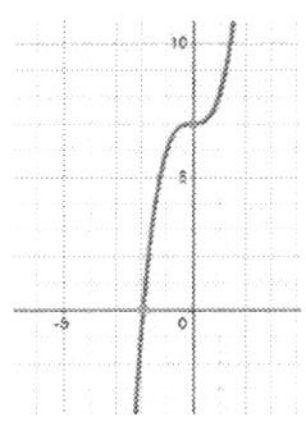

6.

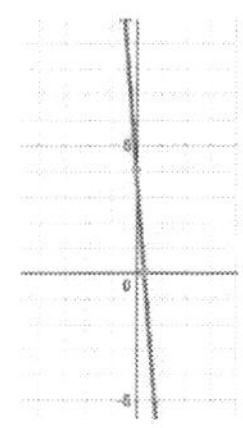

7.

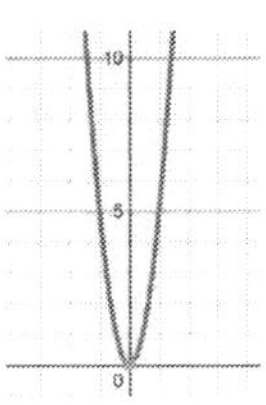

8.

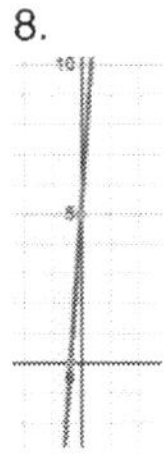

9.

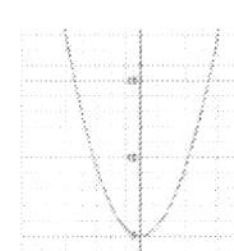

10.

11.

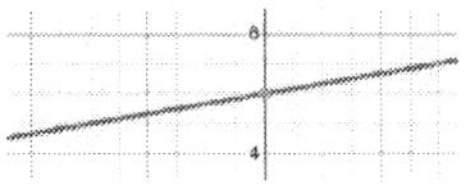

12.

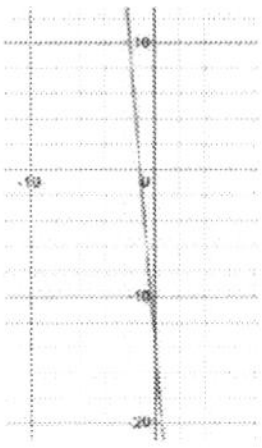

13.

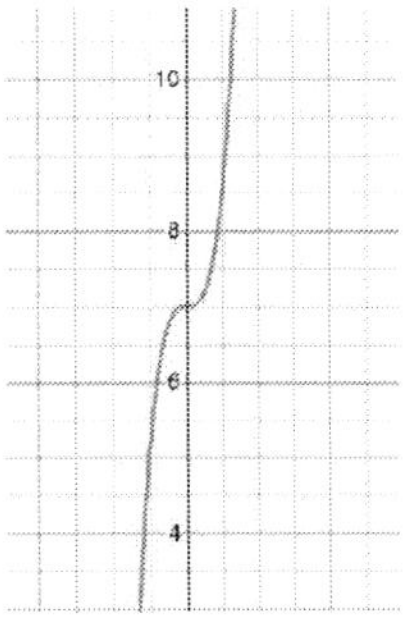

14.

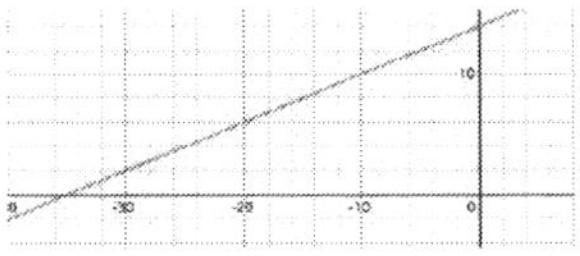

15.

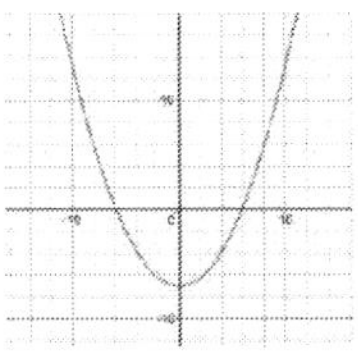

Section 2: Angles and Arcs: Angles in Standard Position

Draw the angles in standard position.

1. 146°

2. 27°

3. 81°

4. 112°

5. 73°

6. 97°

7. 185°

8. 103°

9. 17°

10. 38°

11. 83°

12. 48°

13. 59°

14. 109°

15. 173°

Section 2: Angles and Arcs: Measurement of Angles

Provide the measurement of each angle.

1. 2. 3.

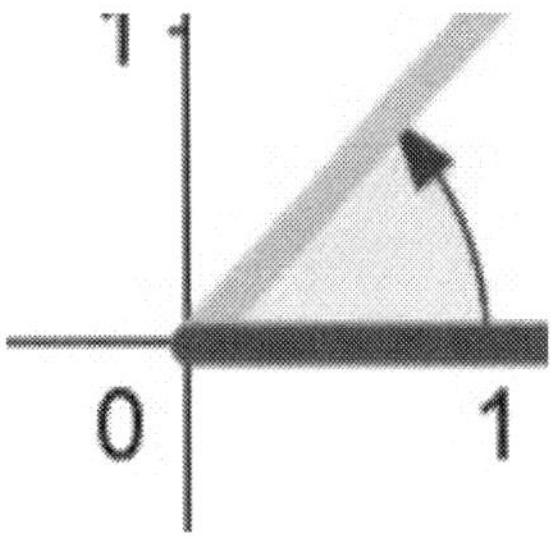

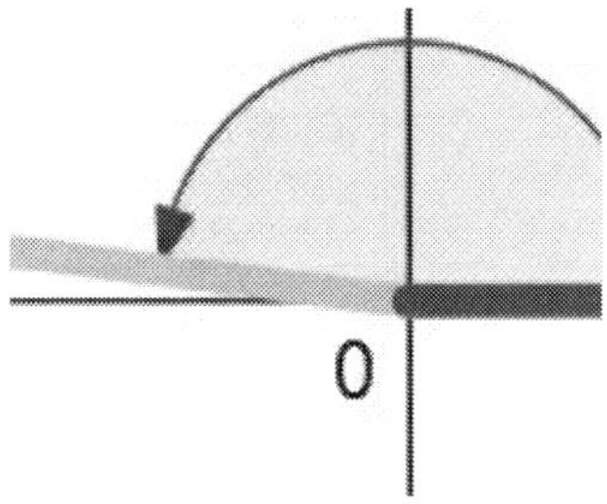

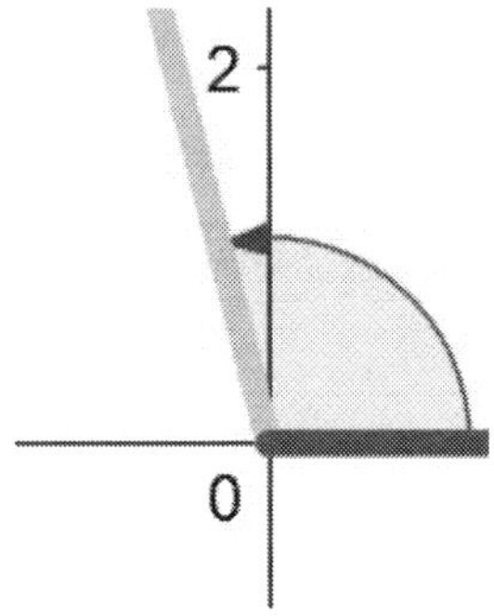

4. 5. 6.

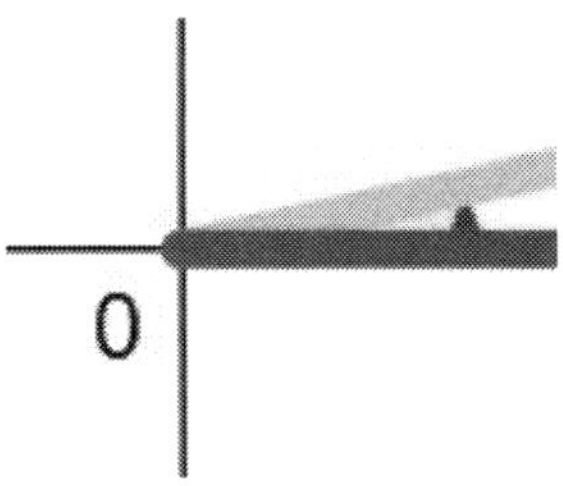

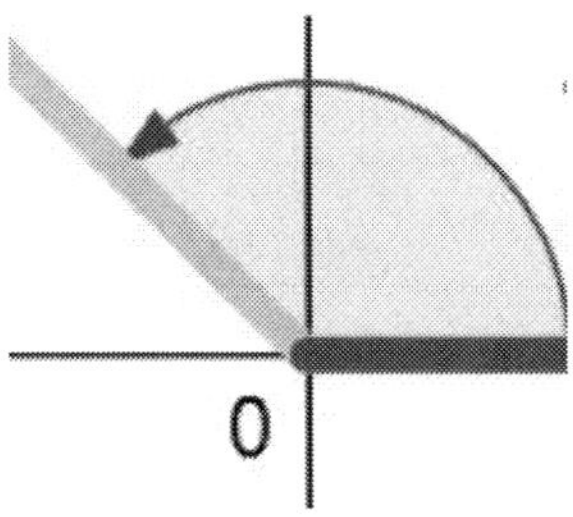

7.

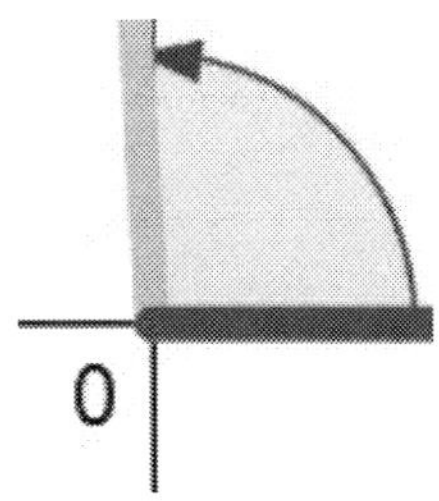

8. 9.

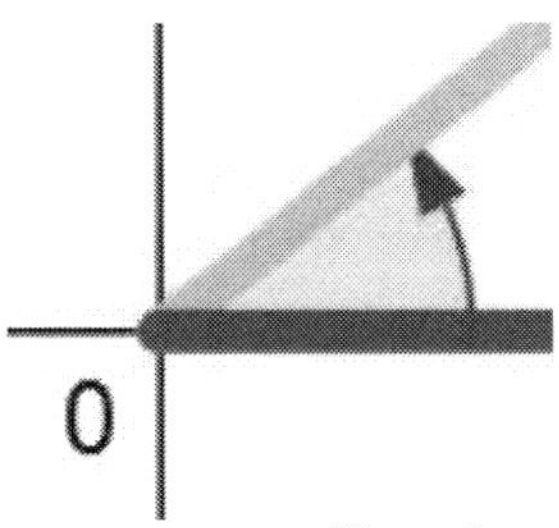

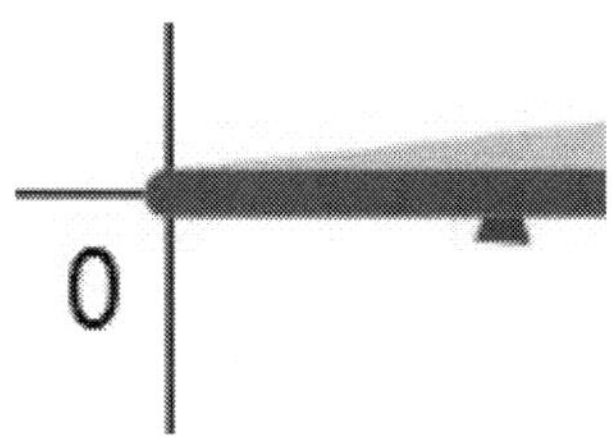

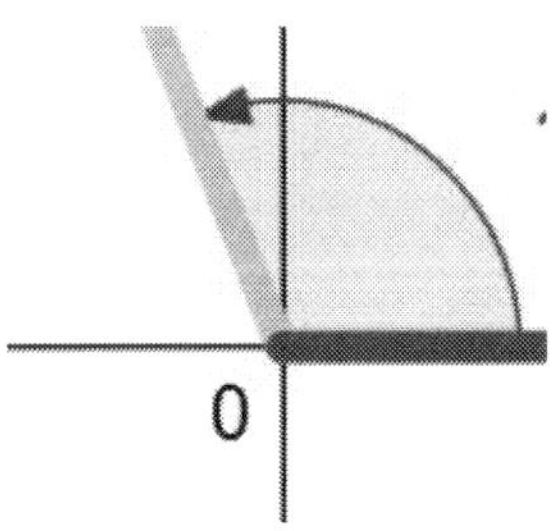

10.

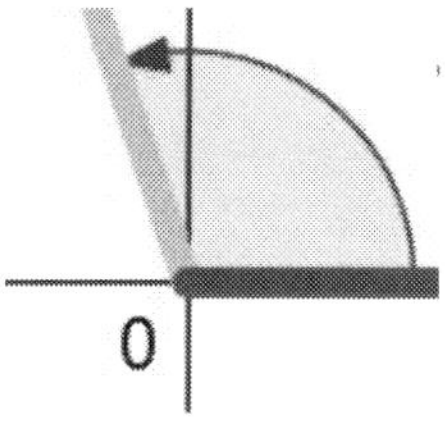

11.

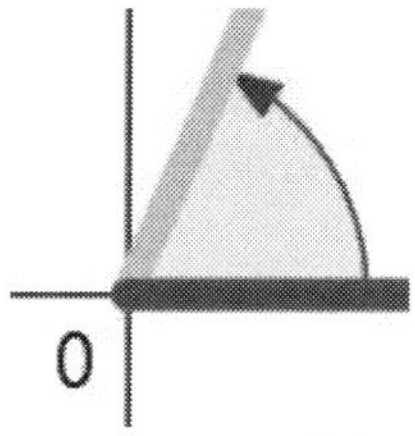

12.

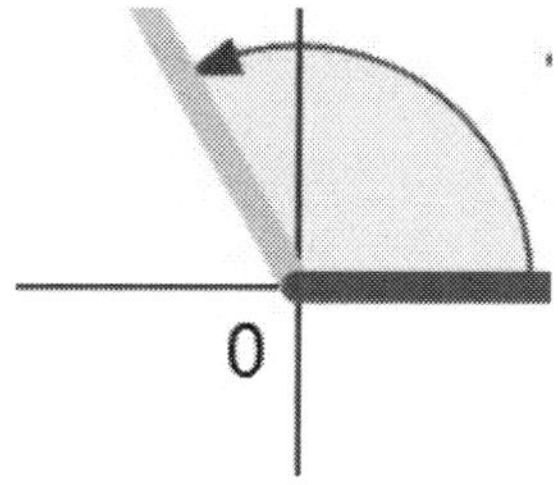

13.

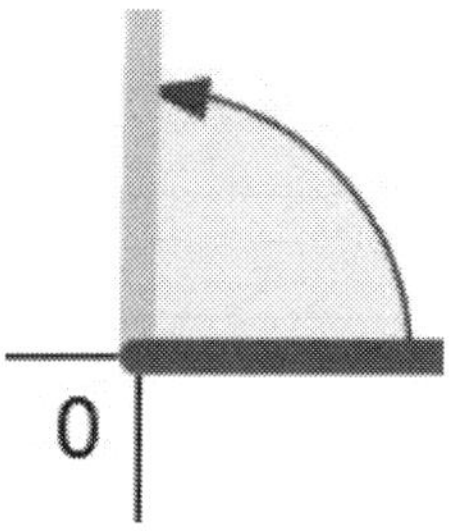

14.

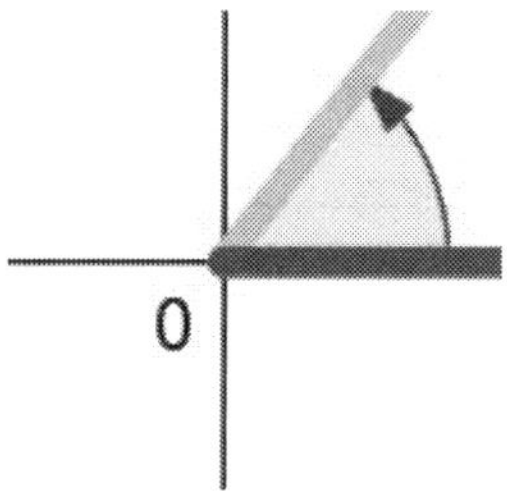

Section 2: Angles and Arcs: Converting Angles

Convert the given degree measurement to radians.

1. $42°$ 2. $117°$ 3. $51°$

4. $89°$ 5. $104°$ 6. $93°$

7. $18°$ 8. $26°$ 9. $35°$

Convert the given radian measurement to degrees.

10. 1.76 radians 11. 0.412 radians 12. 1.21 radians

13. 0.785 radians 14. 0.121 radians 15. 1.49 radians

Section 3: Right Triangle Trigonometry: Pythagorean Theorem

Find the measurement of the missing sides.

1.

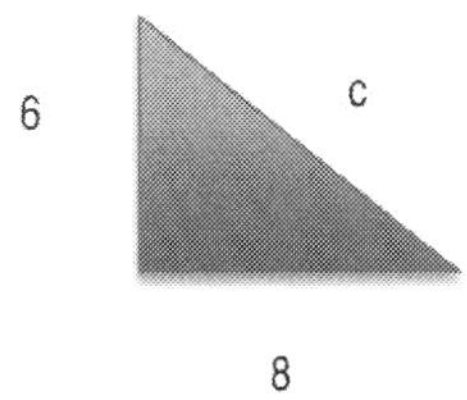

2.

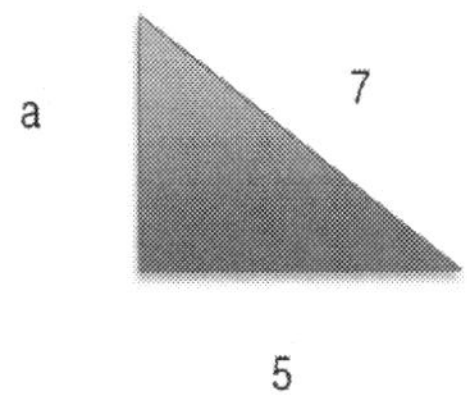

3.

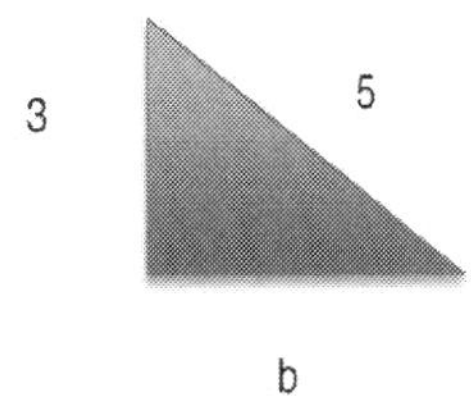

4.

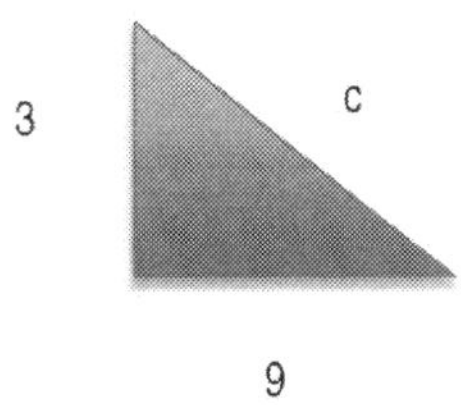

5.

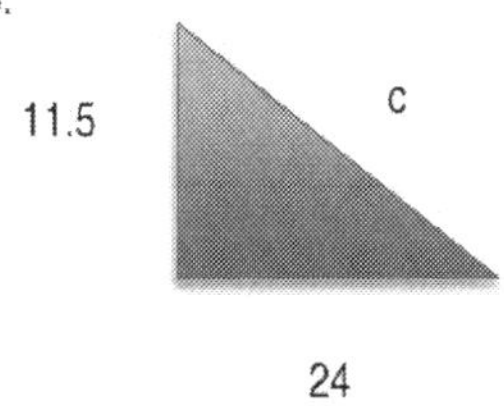

6.

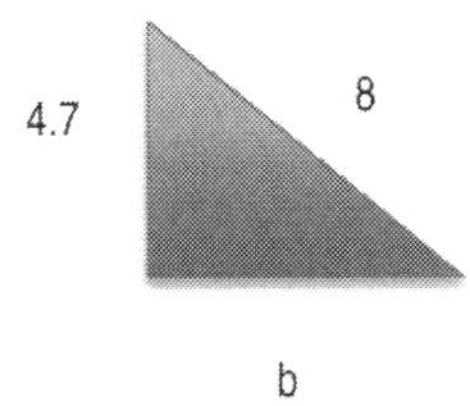

7.

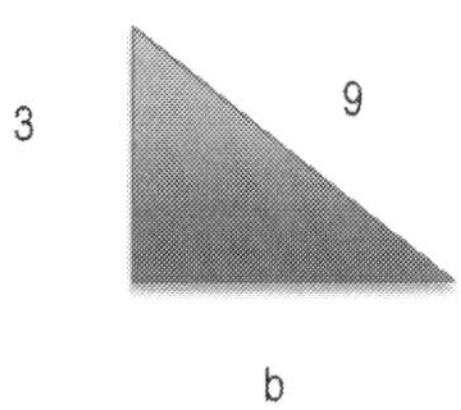

8.

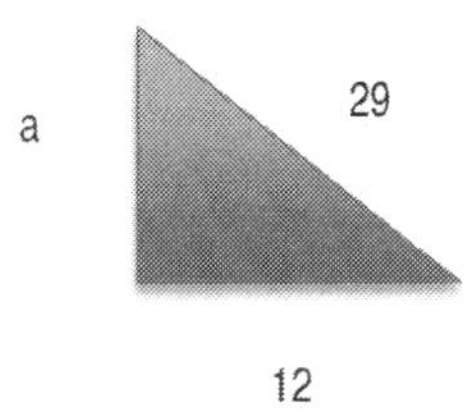

9.

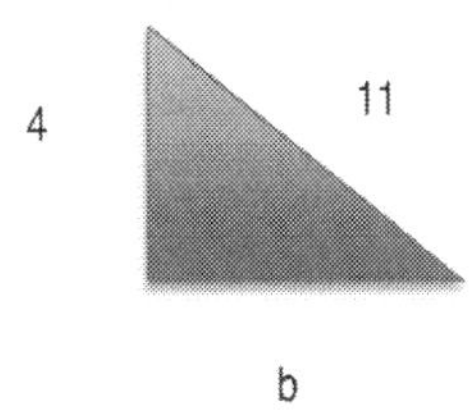

10.

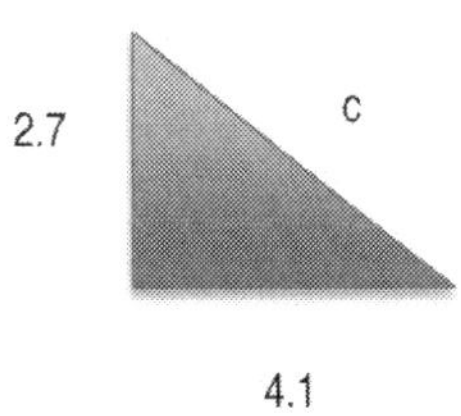

11.

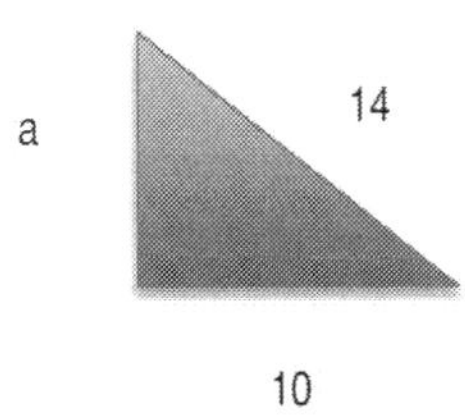

12.

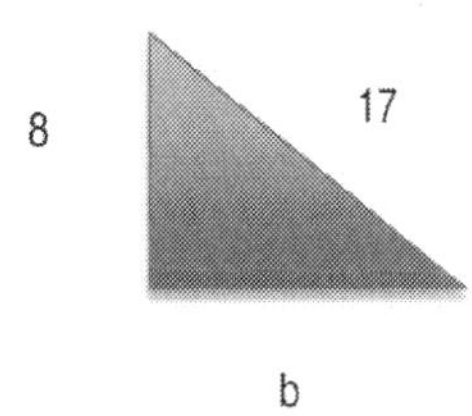

13.

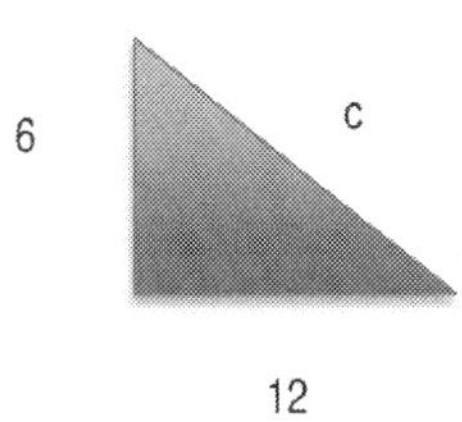

14.

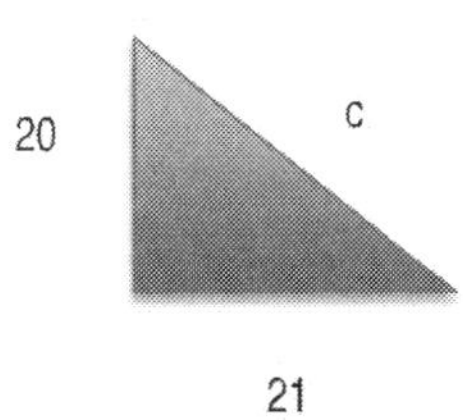

15.

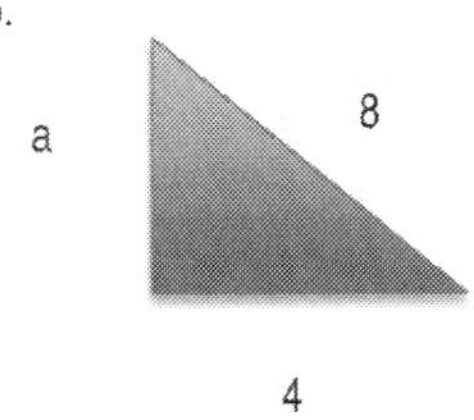

Section 3: Right Triangle Trigonometry: The Six Trigonometric Functions

Use the information provided to find the missing side.

1.

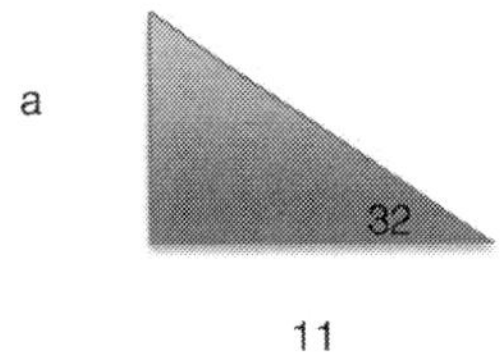

2.

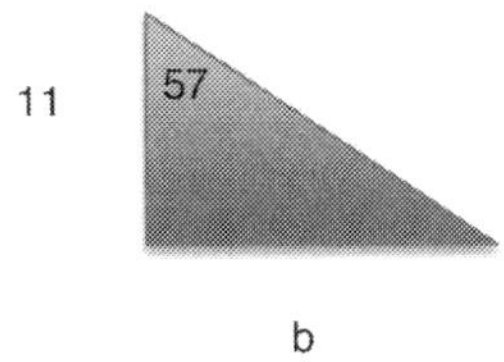

3.

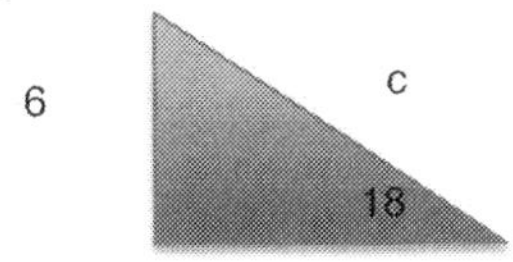

4.

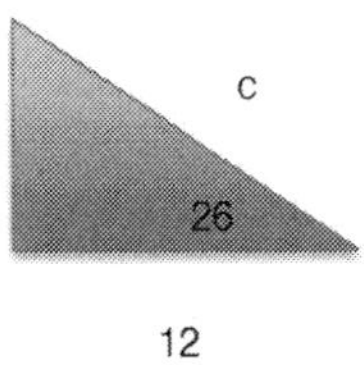

5.

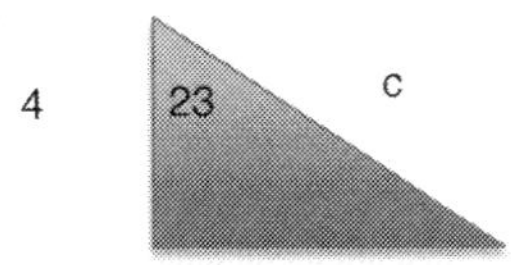

6.

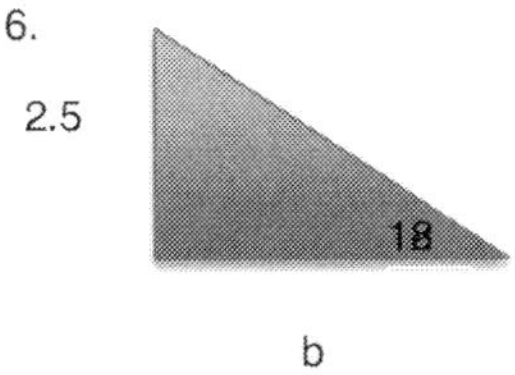

7.

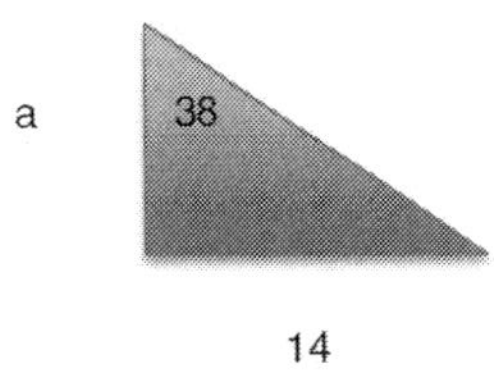

8.

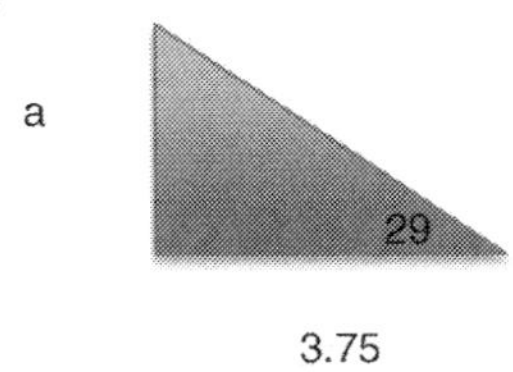

9.

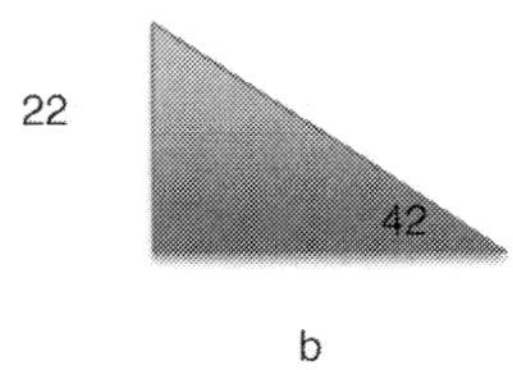

10.

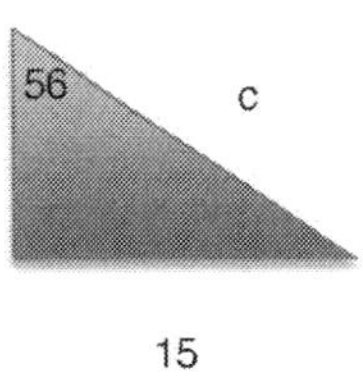

11.

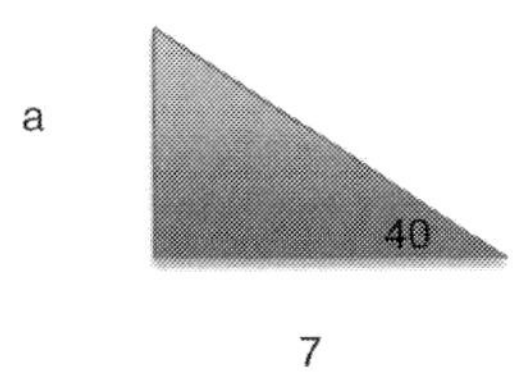

12.

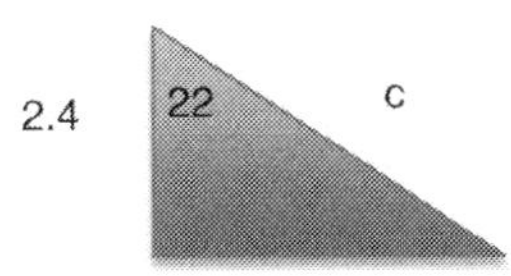

13.

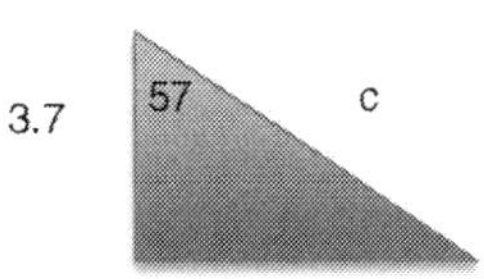

14.

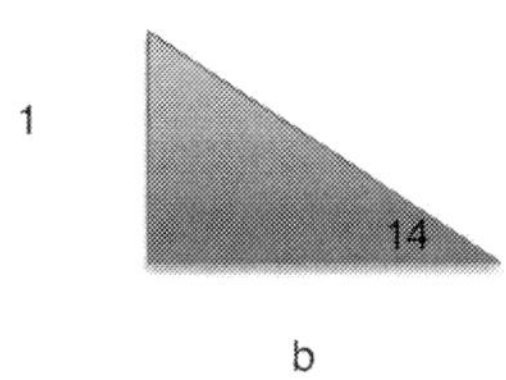

Section 3: Right Triangle Trigonometry: Two Special Right Triangles

1. Find the hypotenuse of an isosceles right triangle with a side length of 4.
2. Find the medium side of a 30, 60, 90 right triangle with a small side of 14.
3. Find the side of an isosceles right triangle with a hypotenuse of 2.5.
4. Find the largest side of a 30, 60, 90 right triangle with a small side of 13.75.
5. Find the hypotenuse of an isosceles right triangle with a side length of 14.
6. Find the side of an isosceles right triangle with a hypotenuse of 9.9.
7. Find the medium side of a 30, 60, 90 right triangle with a small side of 12.6.
8. Find the side of an isosceles right triangle with a hypotenuse of 17.
9. Find the largest side of a 30, 60, 90 right triangle with a small side of 50.
10. Find the largest side of a 30, 60, 90 right triangle with a small side of 75.
11. Find the hypotenuse of an isosceles right triangle with a side length of 11.
12. Find the side of an isosceles right triangle with a hypotenuse of 21.9.
13. Find the largest side of a 30, 60, 90 right triangle with a small side of 1.89.
14. Find the side of an isosceles right triangle with a hypotenuse of 41.
15. Find the hypotenuse of an isosceles right triangle with a side length of 26.

Section 3: Right Triangle Trigonometry: Applying and Solving Right Triangles

1. You calculate a tree to be 75 feet tall as you view it from a 30°degree angle. How far away are you?
2. You are looking at a tree that is 16 feet away from you. The angle of elevation of your viewpoint is 62°. How tall is the tree?
3. You are looking at a tree that is 45 feet away from you. The angle of elevation of your viewpoint is 45°. How tall is the tree?
4. You calculate a tree to be 96 feet tall as you view it from a 51° degree angle. How far away are you?
5. You calculate a tree to be 22 feet tall as you view it from an 82°degree angle. How far away are you?
6. You are looking at a tree that is 16 feet away from you. The angle of elevation of your viewpoint is 12°. How tall is the tree?
7. You calculate a tree to be 142 feet tall as you view it from a 30°degree angle. How far away are you?
8. You are looking at a tree that is 64 feet away from you. The angle of elevation of your viewpoint is 21°. How tall is the tree?

9. You are looking at a tree that is 29 feet away from you. The angle of elevation of your viewpoint is 66°. How tall is the tree?
10. You calculate a tree to be 107 feet tall as you view it from a 23° degree angle. How far away are you?
11. You are looking at a tree that is 52 feet away from you. The angle of elevation of your viewpoint is 60°. How tall is the tree?

12. You calculate a tree to be 100 feet tall as you view it from a 35°degree angle. How far away are you?
13. You are looking at a tree that is 27 feet away from you. The angle of elevation of your viewpoint is 37°. How tall is the tree?
14. You calculate a tree to be 47 feet tall as you view it from a 60°degree angle. How far away are you?
15. You calculate a tree to be 81 feet tall as you view it from a 15°degree angle. How far away are you?

Section 4: The Unit Circle: Using the Unit Circle

1. What is the sine of the angle below? 2. What is the
cosine of the angle below?

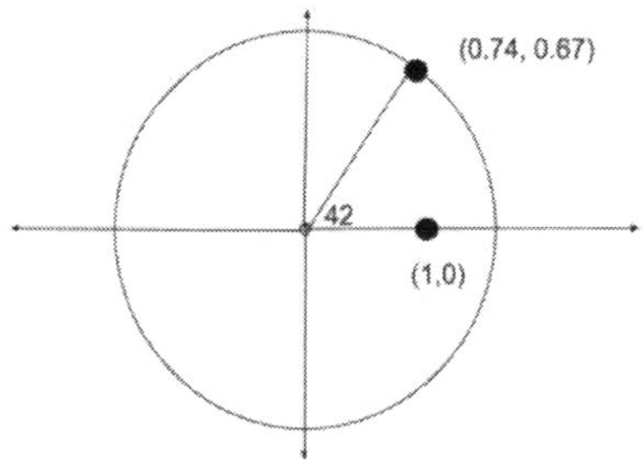

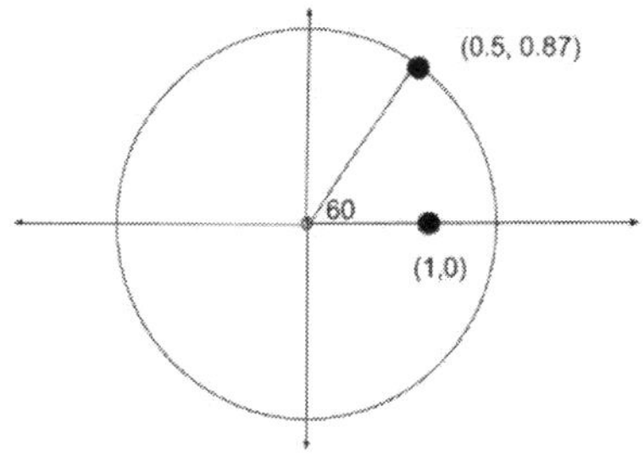

3. What is the cosine of the angle below? 4. What is
the sine of the angle below?

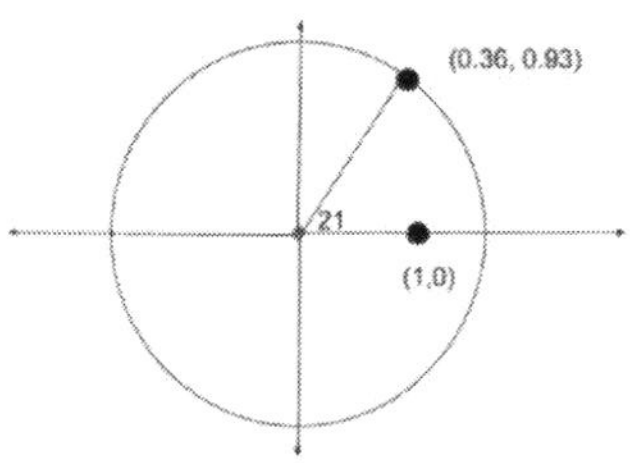

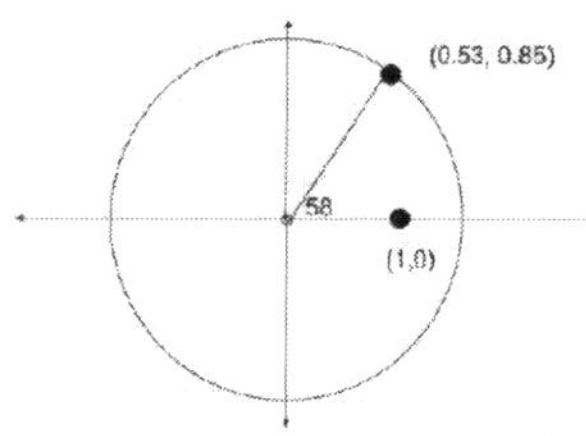

5. What is the sine of the angle below? 6. What is the
cosine of the angle below?

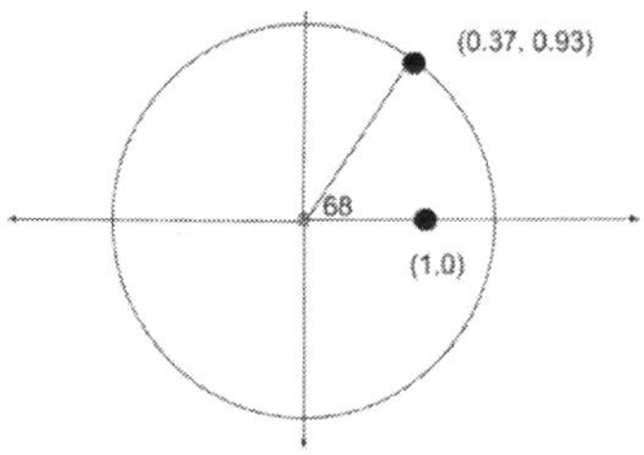

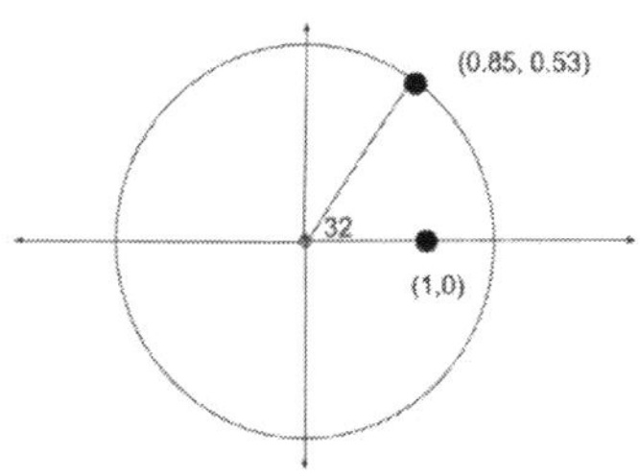

7. What is the angle below? 8. What is
the angle below?

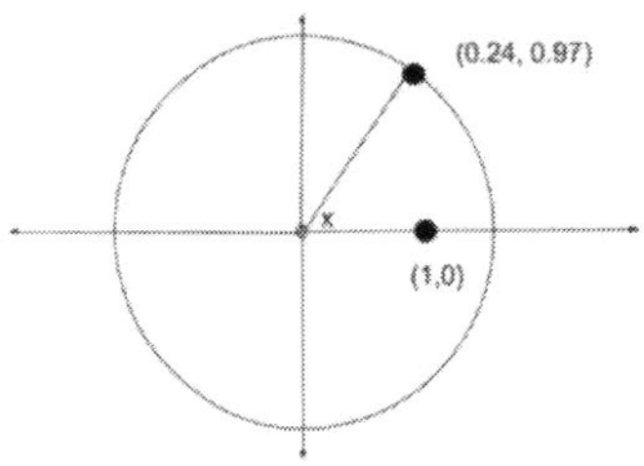

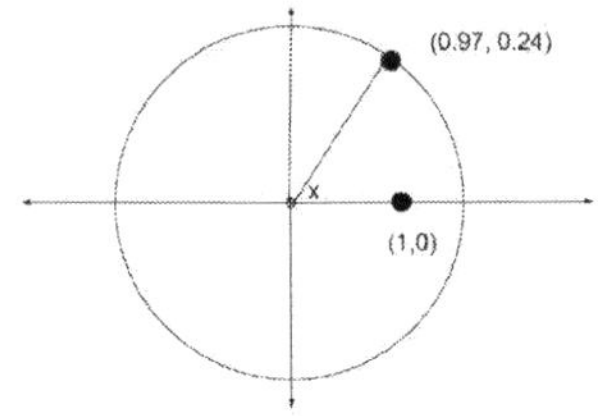

9. What is the angle below? 10. What is
the angle below?

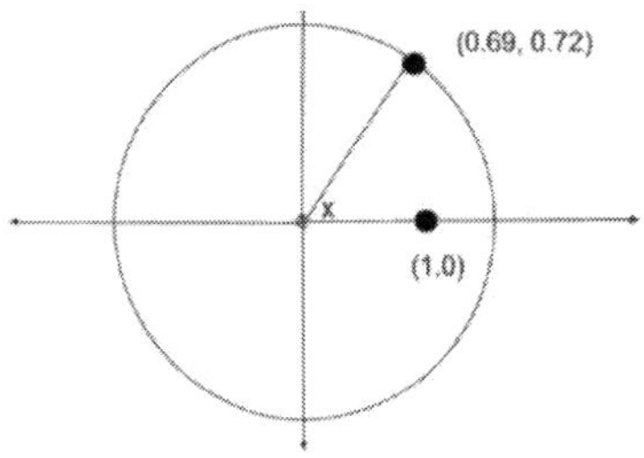

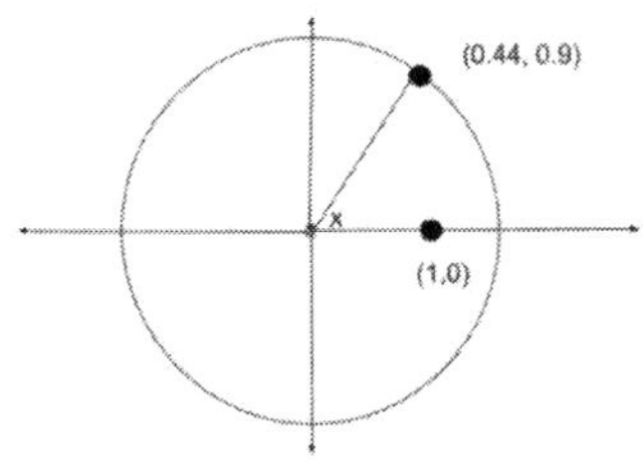

11. What is the sine of the angle below? 12. What is the
cosine of the angle below?

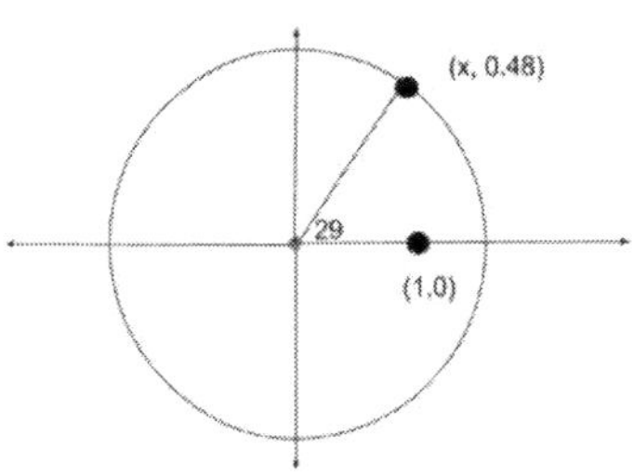

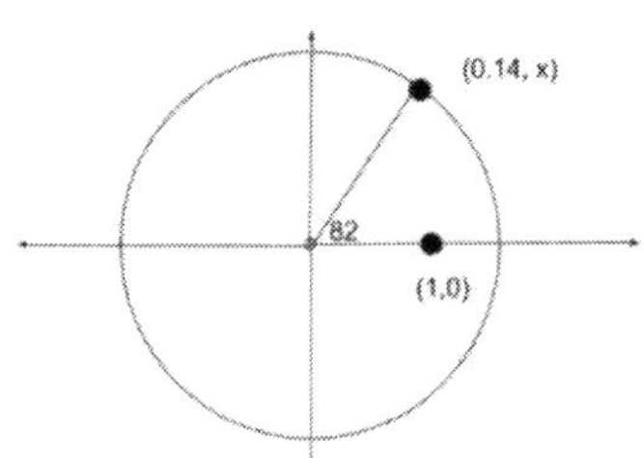

13. What is the cosine of the angle below?

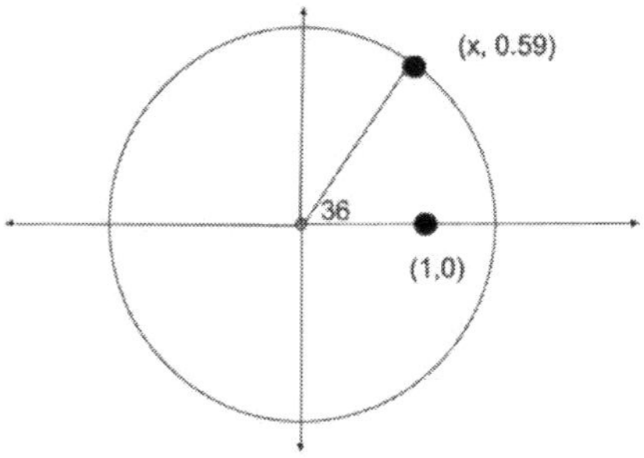

14. What is the sine of the angle below?

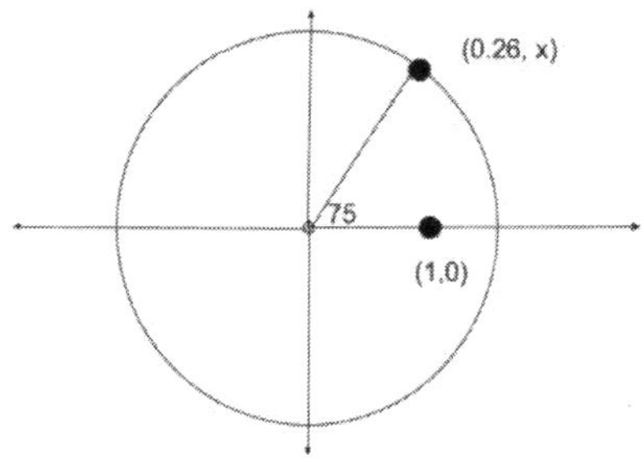

15. What is the angle below?

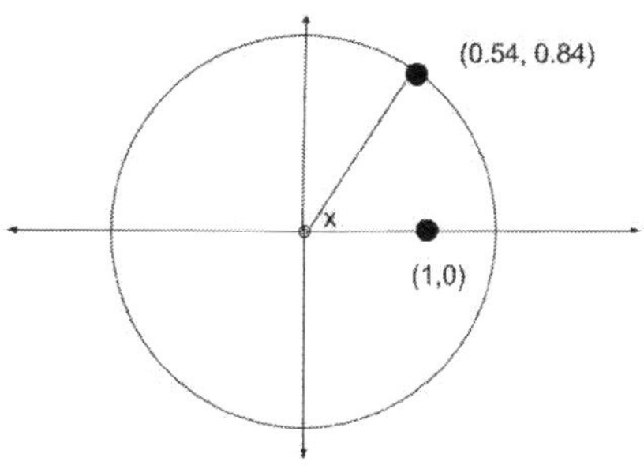

Section 4: The Unit Circle: Pythagorean Identities

Use the information provided, to find the missing identity.

1. If θ is located in Quadrant 1, and sine θ is 0.731, what is the cosine?

2. If θ is located in Quadrant 1, and cosecant θ is 1.008, what is the cotangent?

3. If θ is located in Quadrant 1, and tangent θ is 0.035, what is the secant?

4. If θ is located in Quadrant 1, and cotangent θ is 1.664, what is the cosecant?

5. If θ is located in Quadrant 1, and cosine θ is 0.47, what is the sine?

6. If θ is located in Quadrant 1, and secant θ is 1.941, what is the tangent?

7. If θ is located in Quadrant 1, and sine θ is 0.259, what is the cosine?

8. If θ is located in Quadrant 1, and sine θ is 0.891, what is the cosine?

9. If θ is located in Quadrant 1, and secant θ is 1.071, what is the tangent?

10. If θ is located in Quadrant 1, and cotangent θ is 1.482, what is the cosecant?

11. If θ is located in Quadrant 1, and cosine θ is 0.342, what is the sine?

12. If θ is located in Quadrant 1, and tangent θ is 0.966, what is the secant?

13. If θ is located in Quadrant 1, and cosecant θ is 1.179, what is the cotangent?

14. If θ is located in Quadrant 1, and cosine θ is 0.961, what is the sine?

15. If θ is located in Quadrant 1, and cotangent θ is 0.325, what is the cosecant?

Section 4: The Unit Circle: Reference Angles

Use the information provided to find the reference angle for each measurement.

1. What is the reference angle for an angle that measures 94°?
2. What is the reference angle for an angle that measures 26°?
3. What is the reference angle for an angle that measures 103°?
4. What is the reference angle for an angle that measures 290°?
5. What is the reference angle for an angle that measures 309°?
6. What is the reference angle for an angle that measures 41°?
7. What is the reference angle for an angle that measures 197°?
8. What is the reference angle for an angle that measures 128°?
9. What is the reference angle for an angle that measures 115°?
10. What is the reference angle for an angle that measures 264°?
11. What is the reference angle for an angle that measures 67°?
12. What is the reference angle for an angle that measures 182°?
13. What is the reference angle for an angle that measures 343°?
14. What is the reference angle for an angle that measures 242°?
15. What is the reference angle for an angle that measures 326°?

Section 4: The Unit Circle: Signs of Trigonometric Values

1. The cosine value of an angle is negative and the sine value is positive. Which choice could be a possible measurement for the angle?

   a. 94°

   b. 21°

   c. 317°

   d. 183°

2. The cotangent and tangent of an angle are positive. Which choice could be a possible measurement for the angle?

   a. 146°

   b. 234°

   c. 92°

   d. 307°

3. The sine and tangent of an angle are negative. Which choice could be a possible measurement for the angle?

   a. 183°

   b. 289°

   c. 30°

   d. 91°

4. There are no negative trigonometric values for an angle. Which choice could be a possible measurement for the angle?

   a. 111°

b. 305°

c. 212°

d. 62°

5. The cotangent for an angle is negative but its cosecant is positive. Which choice could be a possible measurement for the angle?

a. 35°

b. 212°

c. 309°

d. 98°

6. The only positive values for an angle are the cosine and secant. Which choice could be a possible measurement for the angle?

a. 342°

b. 200°

c. 100°

d. 50°

7. The sine and cosine for an angle are negative. Which choice could be a possible measurement for the angle?

a. 45°

b. 205°

c. 105°

d. 345°

8. The cosine for an angle is positive and the sine is negative. Which choice could be a possible measurement for the angle?

a. 181°

b. 241°

c. 41°

d. 341°

9. There are no negative trigonometric values for an angle. Which choice could be a possible measurement for the angle?

a. 264°

b. 122°

c. 30°

d. 342°

10. The tangent and cotangent are an angle's only positive values. Which choice could be a possible measurement for the angle?

a. 222°

b. 111°

c. 66°

d. 333°

11. The secant of an angle is positive but its cosecant is negative. Which choice could be a possible measurement for the angle?

a. 75°

b. 275°

c. 175°

d. 25°

12. The tangent and cotangent of an angle are negative. Which choice is not a possible measurement for the angle?

a. 164°

b. 46°

c. 305°

d. 98°

13. The sine and cosine of an angle are positive. Which choice is not a possible measurement for the angle?

a. 16°

b. 21°

c. 96°

d. 62°

14. The cosine of an angle is negative. Which choice is not a possible measurement for the angle?

a. 102°

b. 182°

c. 232°

d. 312°

15. The tangent of an angle is positive. Which choice is not a possible measurement for the angle?

a. 95°

b. 185°

c. 45°

d. 205°

Section 5: Graphing and Inverse Trigonometric Functions: Graphing Trigonometric Functions

Review your knowledge of trigonometric functions and identify each graph pictured.

1.

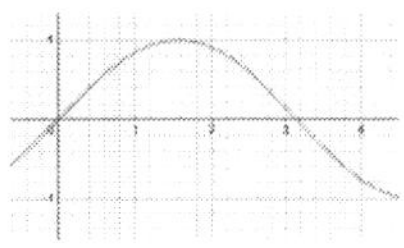

2.

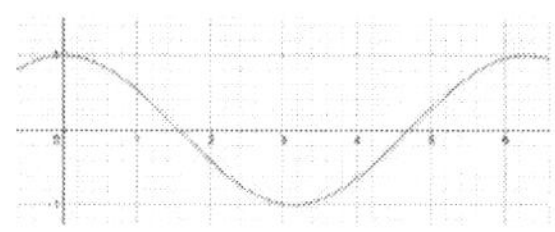

3.

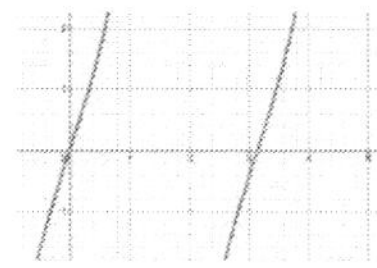

4.

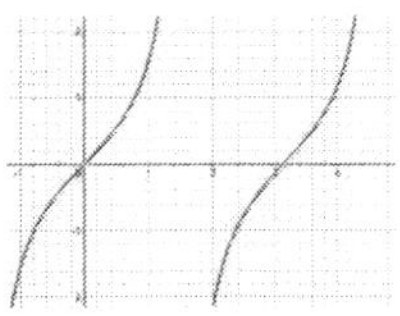

5.

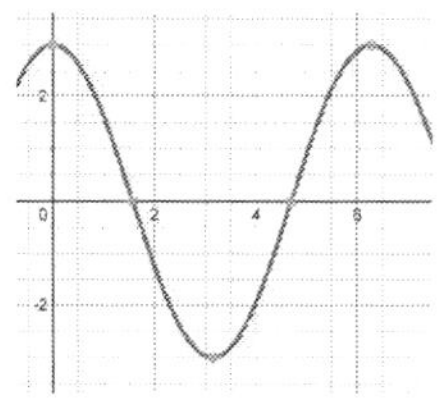

6.

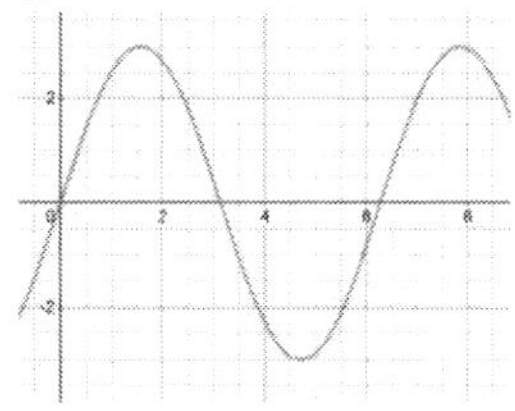

7.

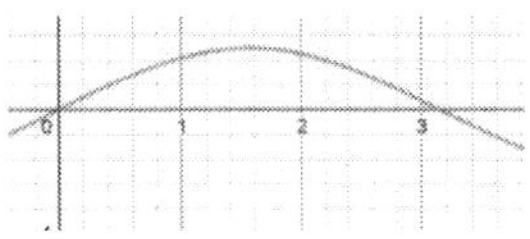

8.

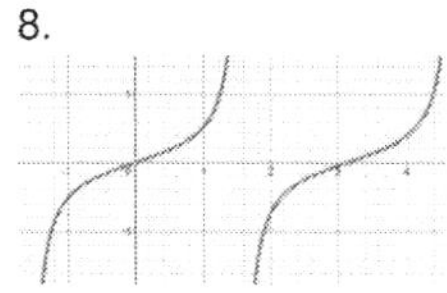

9.

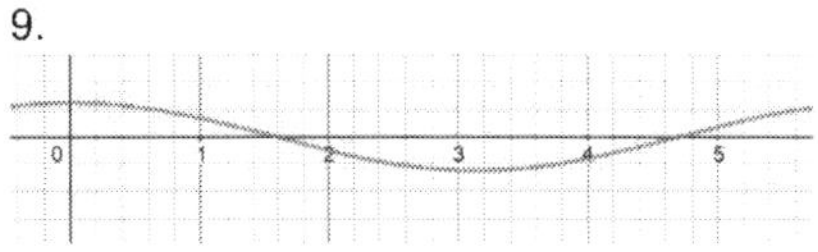

10.

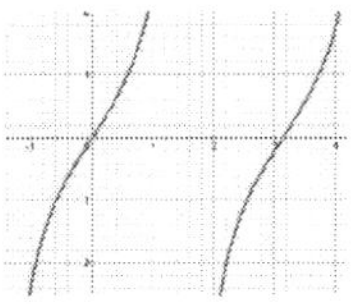

11.

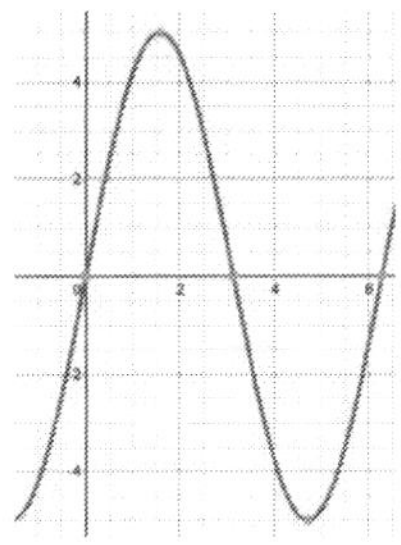

12.

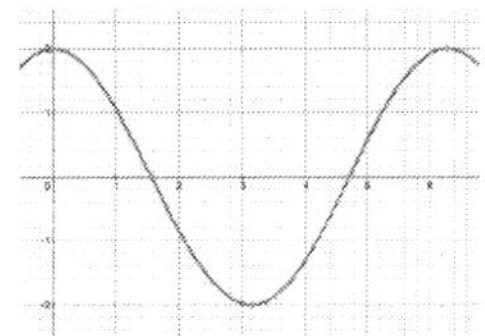

13.

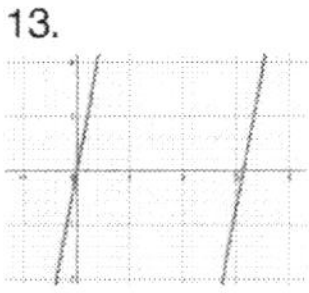

14.

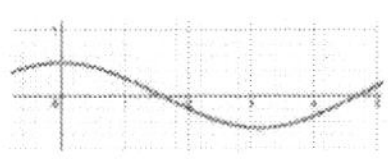

15.

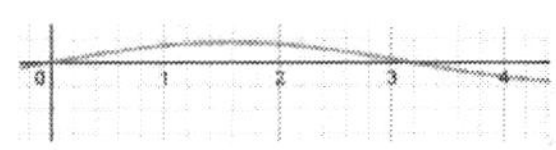

Section 5: Graphing and Inverse Trigonometric Functions: Properties of Trigonometric Functions

Find the information requested for each provided graph.

Graph 1

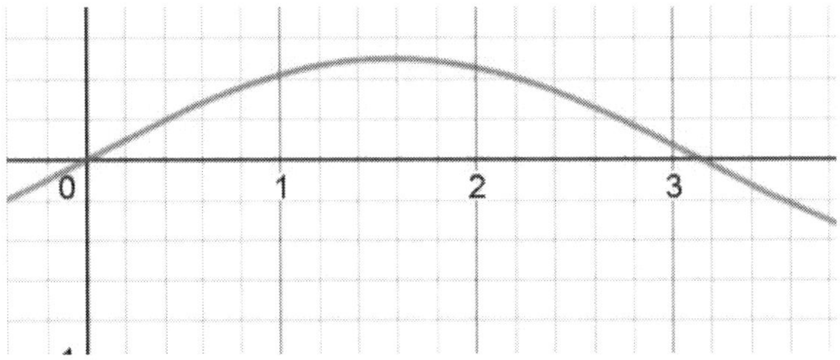

1. Period
2. Amplitude
3. Domain

Graph 2

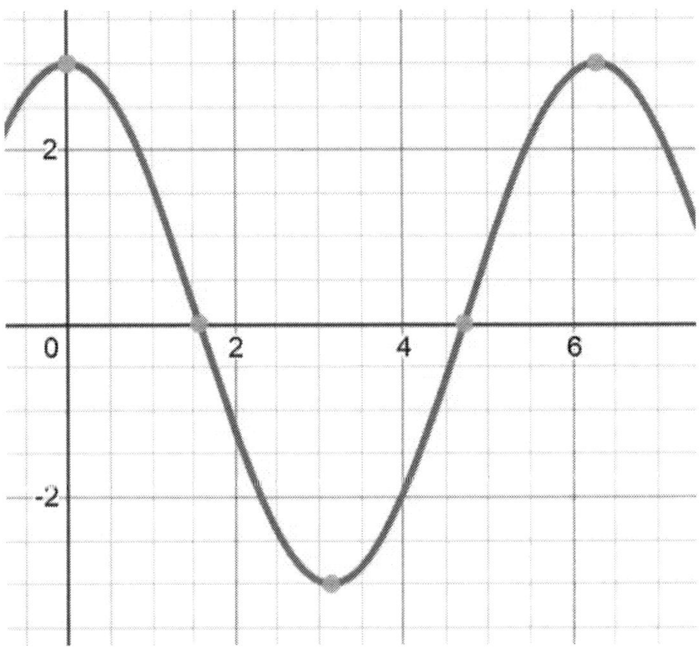

1. Range 2. Domain 3. Amplitude

Graph 3

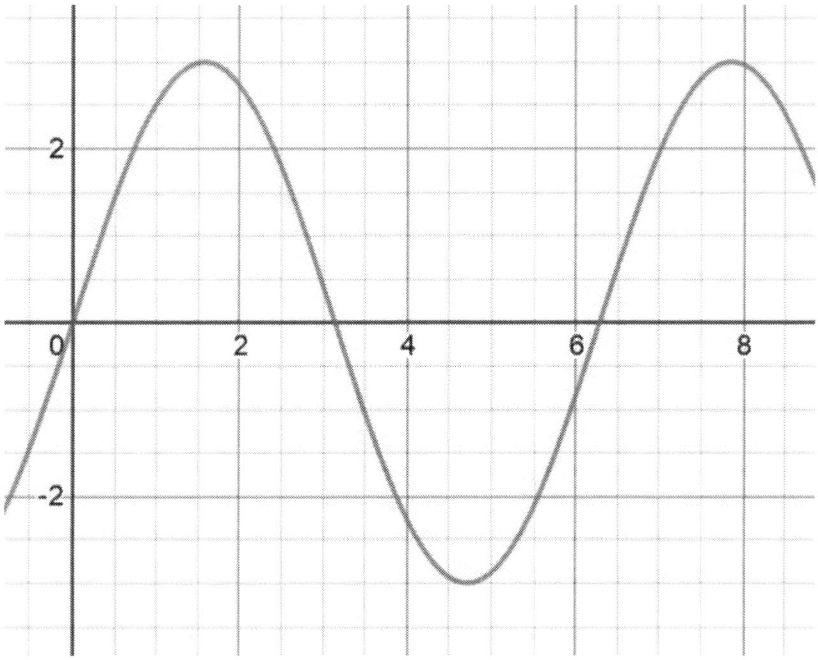

1. Period 2. Range 3. Amplitude

Graph 4

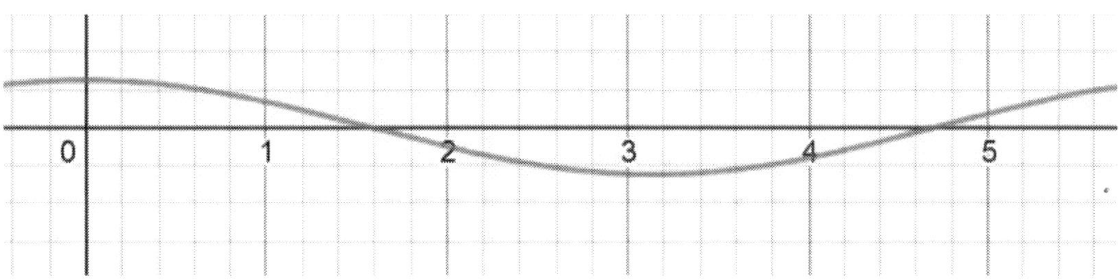

1. Range 2. Domain 3. Period

Graph 5

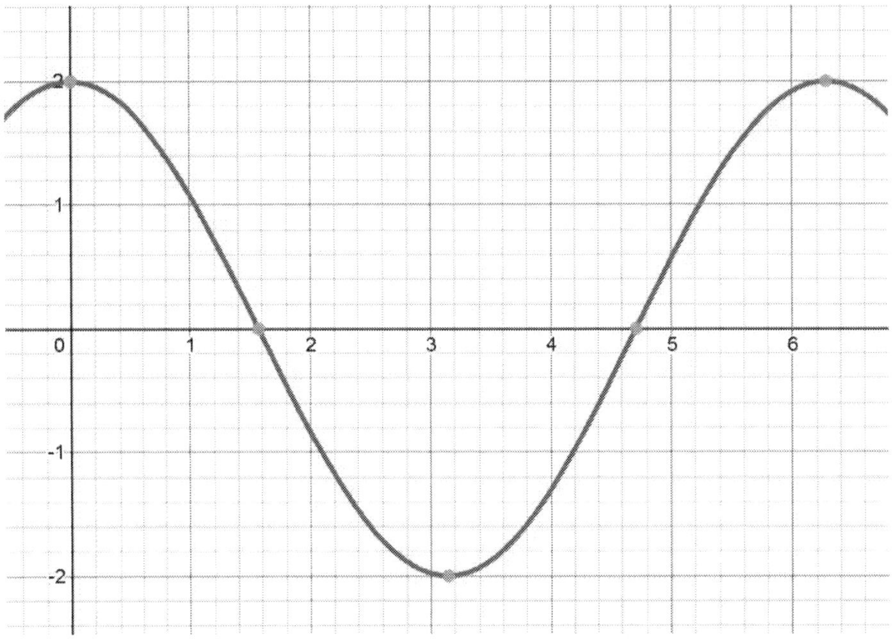

1. Range 2. Domain 3. Amplitude

Section 5: Graphing and Inverse Trigonometric Functions: Inverse Trigonometric Functions

What type of function is shown by each graph provided?

1.

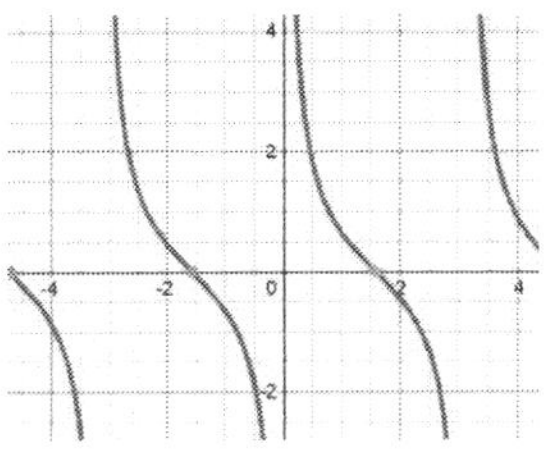

2.

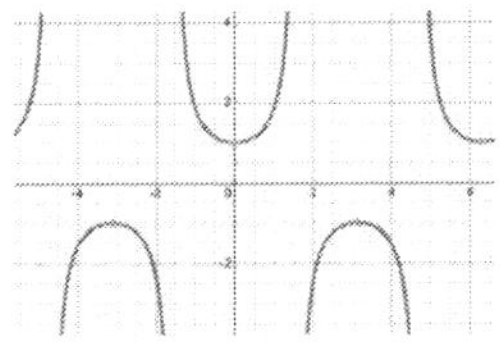

3.

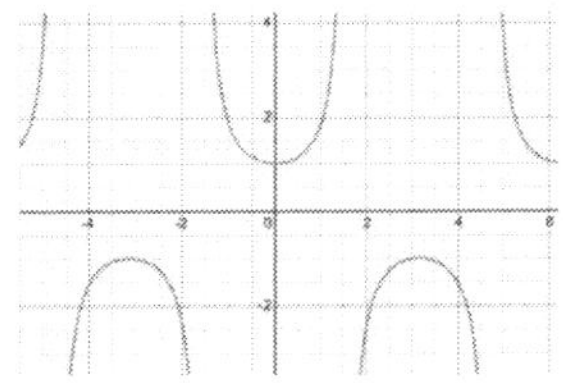

4.

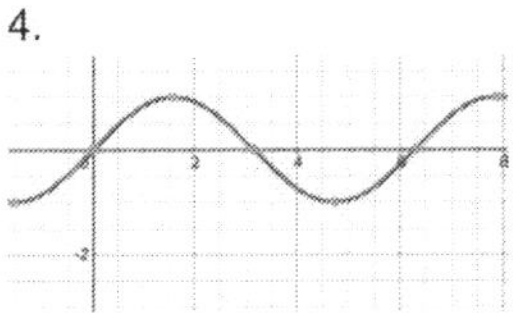

5.

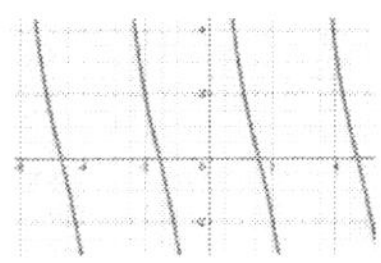

6.

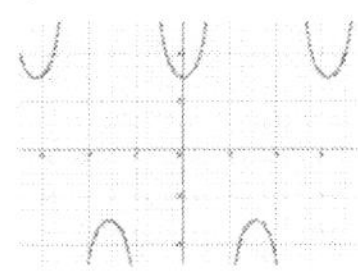

7.

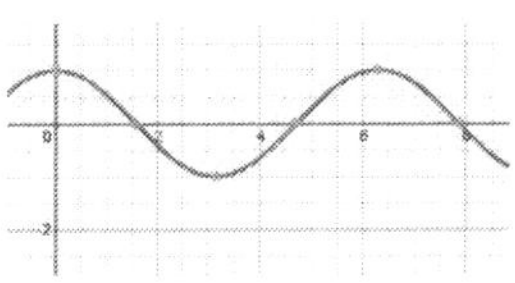

8.

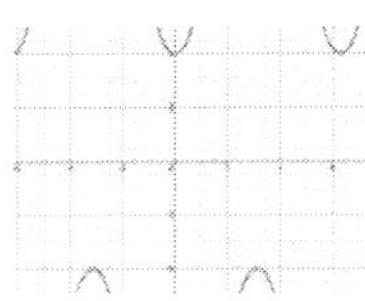

9.

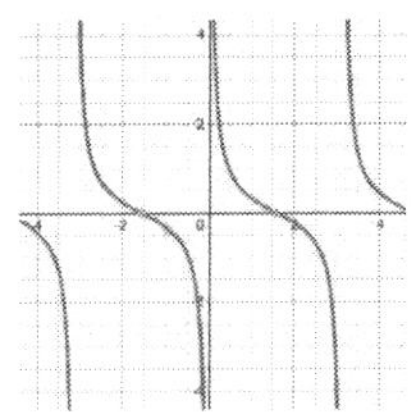

10.

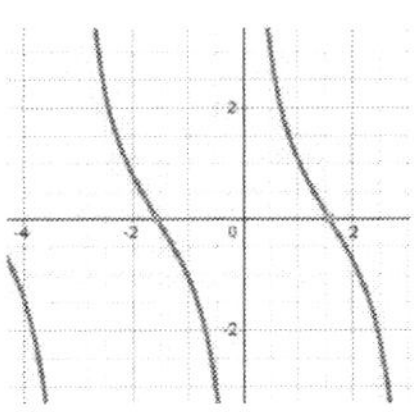

11.

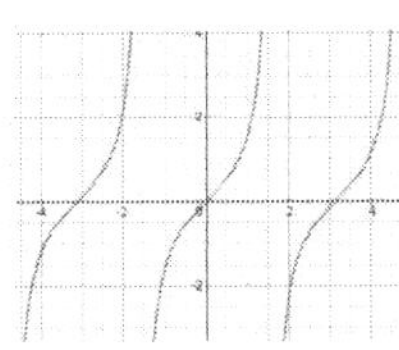

12.

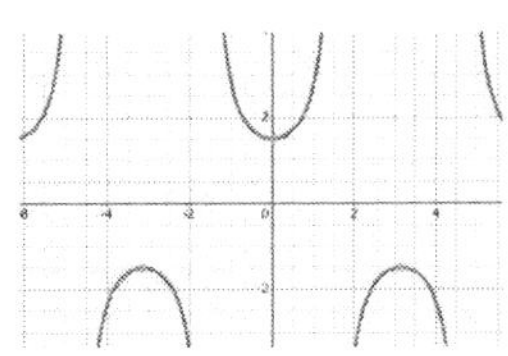

Section 6: Trigonometric Identities: Proving Identities

1. Prove $1 + \cot^2 A = \csc^2 A$ using an angle of 78°, showing each step.

2. Prove $\sin^2 A + \cos^2 A = 1$ using an angle of 60°, showing each step.

3. Prove $\sin^2 A + \cos^2 A = 1$ using an angle of 32°, showing each step.

4. Prove $\tan^2 A + 1 = \sec^2 A$ using an angle of 86°, showing each step.

5. Prove $\tan^2 A + 1 = \sec^2 A$ using an angle of 52°, showing each step.

6. Prove $1 + \cot^2 A = \csc^2 A$ using an angle of 13°, showing each step.

7. Prove $\sin^2 A + \cos^2 A = 1$ using an angle of 29°, showing each step.

8. Prove $\tan^2 A + 1 = \sec^2 A$ using an angle of 69°, showing each step.

9. Prove $1 + \cot^2 A = \csc^2 A$ using an angle of 46°, showing each step.

A gift for you

Enjoy your gift.

10. Prove $\sin^2 A + \cos^2 A = 1$ using an angle of 58°, showing each step.

11. Prove $\tan^2 A + 1 = \sec^2 A$ using an angle of 12°, showing each step.

12. Prove $1 + \cot^2 A = \csc^2 A$ using an angle of 82°, showing each step.

13. Prove $\tan^2 A + 1 = \sec^2 A$ using an angle of 45°, showing each step.
14. Prove $\sin^2 A + \cos^2 A = 1$ using an angle of 51°, showing each step.

15. Prove $1 + \cot^2 A = \csc^2 A$ using an angle of 35°, showing each step.

Section 7: Trigonometric Equations: Trigonometric Equations
Solve each equation for x.
1. $5 + \cos x = 6$

2. $10 - \tan x = 9$

3. $17 + \sin x = 18$

4. $21 - \sin x = 22$

5. $14 - \cos x = 15$

6. $3 \cos x = 3$

7. $10 \sin x = 10$

8. $5 \tan x = 5$

9. $30 \sin x = 15$

10. $26 \cos x = 13$

11. $15 \cos x = -15$

12. $12 \tan x = 0$

13. $32\left(\frac{1}{2} \sin x\right) = 8$

14. $48\left(\frac{1}{4} \cos x\right) = 6$

15. $-4(2 \sin x) = 8$

Section 8: Oblique Triangle Laws: The Law of Sines

1. Angle A measures 89° and angle B measures 26°. Side b measures 1.76. What does side a measure?

2. Angle A measures 13° and angle B measures 35°. Side a measures 14.1. What does side b measure?

3. Angle A measures 83° and angle B measures 11°. Side b measures 9.4. What does side a measure?

4. Angle A measures 50° and angle B measures 27°. Side b measures 22.5. What does side a measure?

5. Angle A measures 39° and angle B measures 40°. Side a measures 17. What does side b measure?

6. Angle A measures 72° and angle B measures 36°. Side a measures 3.89. What does side b measure?

7. Angle A measures 39° and angle B measures 81°. Side b measures 6.25. What does side a measure?

8. Angle A measures 67° and angle B measures 12°. Side a measures 9.75. What does side b measure?

9. Angle A measures 46° and angle B measures 60°. Side b measures 11.6. What does side a measure?

10. Angle A measures 65° and angle B measures 48°. Side a measures 21.17. What does side b measure?
11. Angle A measures 9° and angle B measures 55°. Side b measures 3.41. What does side a measure?
12. Angle A measures 42° and angle B measures 59°. Side b measures 16.2. What does side a measure?

13. Angle A measures 90° and angle B measures 48°. Side a measures 28.7. What does side b measure?

14. Angle A measures 26° and angle B measures 78°. Side b measures 50. What does side a measure?

15. Angle A measures 52° and angle B measures 7°. Side a measures 147. What does side b measure?

Section 8: Oblique Triangle Laws: The Law of Cosines

1. Side a of a triangle measures 4 units and side b measures 7 units. Angle C measures 34°. What is the measurement of side c?

2. Side b of a triangle measures 2.6 units and side c measures 4.2 units. Angle C measures 61°. What is the measurement of side a?

3. Side a of a triangle measures 13 units and side c measures 19 units. Angle C measures 21°. What is the measurement of side b?

4. Side a of a triangle measures 12 units and side b measures 16 units. Angle C measures 68°. What is the measurement of side c?

5. Side b of a triangle measures 113 units and side c measures 121 units. Angle C measures 48°. What is the measurement of side a?

6. Side a of a triangle measures 3. Side b measures 4 and side c measures 5. What is the measurement of angle C?

7. Side a of a triangle measures 17 units and side b measures 22 units. Angle C measures 25°. What is the measurement of side c?

8. Side a of a triangle measures 3 units and side c measures 5 units. Angle C measures 19°. What is the measurement of side b?

9. Side a of a triangle measures 26 units and side c measures 28 units. Angle C measures 86°. What is the measurement of side b?

10. Side a of a triangle measures 8 units and side b measures 6 units. Angle C measures 90°. What is the measurement of side c?
11. Side b of a triangle measures 15 units and side c measures 20 units. Angle C measures 75°. What is the measurement of side a?
12. Side a of a triangle measures 39 units and side c measures 39 units. Angle C measures 41°. What is the measurement of side b?

13. Side a of a triangle measures 1.61 units and side b measures 2.31 units. Angle C measures 42°. What is the measurement of side c?

14. Side a of a triangle measures 16. Side b measures 21 and side c measures 32. What is the measurement of angle C?

15. Side a of a triangle measures 11 units and side c measures 21 units. Angle C measures 112°. What is the measurement of side b?

Section 9: Complex Numbers and Polar Coordinates: Plotting Imaginary Numbers

How would you plot these imaginary numbers?

1. 10+9i
2. 4-7i
3. -15+2i

4. 6+5i
5. -9-10i
6. 28+2i

7. -1-10i
8. -19+3i
9. 13-i

10. 14- 13i
11. 21+ 15i
12. -2-6i

13. -11+i
14. 5+11i
15. 7-8i

Section 9: Complex Numbers and Polar Coordinates: Combining Complex Numbers

Combine these complex numbers.

1. (6+i) +(12+7i) 2. (10-11i)-(9- 16i) 3. (3-6i)-(1-14i)

4. (9-7i)+(13-15i) 5. (1+2i)-(5+17i) 6. (8-12i)+(4+6i)

7. (12-3i)-(16-12i) 8. (6+10i)+(7+ 4i) 9. (18-5i)-(18-13i)

10. (5+ 8i)+(10 +3i) 11. (13 +20i)+ (21-11i) 12. (4-13i)-(14+8i)

13. (7+ 18i)-(15-2i) 14. (25 -4i)+(16+10i)
15. (9+9i)-(27-i)

Section 9: Complex Numbers and Polar Coordinates: Multiplying Complex Numbers

Multiply these complex numbers.

1. (11+ 5i)(5-10i) 2. (6-4i)(6+9i) 3. (10-5i)(11-2i)

4. (5-12i)(10-3i) 5. (6-i)(5+7i) 6. (9+2i)(3+2i)

7. (20-7i)(8-8i) 8. (12-10i)(1+4i) 9. (4-20i)(6-i)

10. (8+ 3i)(20-5i) 11. (3-2i)(7+i) 12. (7+6i)(2+6i)

13. (2-11i)(3-4i)

14. (7+9i)(4-2i)

15. (1-3i)(9-2i)

Section 9: Complex Numbers and Polar Coordinates: Complex Numbers in Polar Form

Rewrite these complex numbers in their polar form.

1. -6+4i 2. 5+5i 3. -2-7i

4. 14+12i 5. -32-16i 6. -4-3i

7. 12-5i 8. 9+3i 9. 5i

10. 3-2i 11. 8+13i 12. 12-10i

13. 10- 3i 14. 27+5i 15. 21-6i

Problem Answers

Section 1: Introducing Trigonometry-The Distance Between Two Points

1. 18.03
2. 28.16
3. 13
4. 8.602
5. 27.46
6. 16.49
7. 19.85
8. 67.42
9. 35.81
10. 43.01
11. 15.26
12. 20.59
13. 40.5
14. 23.71
15. 5

Section 1: Introducing Trigonometry- Functions

1. Function; +3
2. Not a function
3. Not a function
4. Function, /12
5. Not a function
6. Function, 5x-7
7. Not a function
8. Not a function
9. Function, x^3
10. Not a function
11. Function, -7
12. Not a function
13. Function, 4x+2
14. Function, x^2
15. Not a function

Section 1: Introducing Trigonometry- Graphs of Functions
1. Linear
2. Quadratic
3. Linear
4. Linear
5. Quadratic
6. Linear
7. Quadratic
8. Linear
9. Quadratic
10. Linear
11. Linear
12. Linear
13. Quadratic
14. Linear
15. Quadratic

Section 2: Angles and Arcs: Angles in Standard Position
1. The angle should measure the degree amount stated, measured from the x-axis.
2. The angle should measure the degree amount stated, measured from the x-axis.
3. The angle should measure the degree amount stated, measured from the x-axis.
4. The angle should measure the degree amount stated, measured from the x-axis.
5. The angle should measure the degree amount stated, measured from the x-axis.
6. The angle should measure the degree amount stated, measured from the x-axis.
7. The angle should measure the degree amount stated, measured from the x-axis.
8. The angle should measure the degree amount stated, measured from the x-axis.
9. The angle should measure the degree amount stated, measured from the x-axis.

10. The angle should measure the degree amount stated, measured from the x-axis.
11. The angle should measure the degree amount stated, measured from the x-axis.
12. The angle should measure the degree amount stated, measured from the x-axis.
13. The angle should measure the degree amount stated, measured from the x-axis.
14. The angle should measure the degree amount stated, measured from the x-axis.
15. The angle should measure the degree amount stated, measured from the x-axis.

Section 2: Angles and Arcs: Measurement of Angles

1. $47°$
2. $173°$
3. $104°$
4. $12°$
5. $136°$
6. $93°$
7. $37°$
8. $6°$
9. $112°$
10. $109°$
11. $66°$
12. $119°$
13. $89°$
14. $52°$

Section 2: Angles and Arcs: Converting Angles

1. 0.733 radians
2. 2.042 radians
3. 0.890 radians
4. 1.553 radians
5. 1.815 radians
6. 1.623 radians

7. 0.314 radians
8. 0.454 radians
9. 0.611 radians
10. $100°$
11. $24°$
12. $69°$
13. $45°$
14. $7°$
15. $85°$

Section 3: Right Triangle Trigonometry: Pythagorean Theorem
1. c=10
2. a=4.9
3. b=4
4. c=9.4
5. c=26.6
6. b=6.4
7. b=8.5
8. a= 26.4
9. b=10.2
10. c=4.9
11. a= 9.8
12.b=15
13. c=13.4
14. c=29
15. a=6.9

Section 3: Right Triangle Trigonometry: The Six Trigonometric Functions
1. a=7
2. b=17
3. c=9.5
4. c=13.4
5. c =9.8
6. b=7.5
7. a=11

8. a=2.1
9. b=24
10. c=18
11. a=6
12. c=6.6
13. b=5.8
14. b=4

Section 3: Right Triangle Trigonometry: Two Special Right Triangles

1. c= 5.7
2. s=24.2
3. s= 1.8
4. c=27.5
5. c=19.8
6. s= 7
7. s=21.8
8. s=12
9. c=100
10. c=150
11. c=15.6
12. s=15.5
13. s=3.3
14. s= 29
15. c=36.8

Section 3: Right Triangle Trigonometry: Applying and Solving Right Triangles

1. 130 feet
2. 30 feet
3. 45 feet
4. 78 feet
5. 3 feet
6. 3 feet
7. 246 feet
8. 25 feet

9. 65 feet
10. 252 feet
11. 90 feet
12. 143 feet
13. 20 feet
14. 27 feet
15. 302 feet

Section 4: The Unit Circle: Using the Unit Circle
1. 0.67
2. 0.5
3. 0.93
4. 0.85
5. 0.93
6. 0.85
7. 76°
8. 14°
9. 46°
10. 64°
11. 0.99
12. 0.87
13. 0.81
14. 0.97
15. 57°

Section 4: The Unit Circle: Pythagorean Identities
1. 0.681
2. 0.123
3. 1.001
4. 1.941
5. 0.882
6. 1.664
7. 0.966
8. 0.459
9. 0.384
10. 1.788

11. 0.94
12. 1.39
13. 0.625
14. 0.276
15. 1.051

Section 4: The Unit Circle: Reference Angles

1. 86°
2. 26°
3. 77°
4. 70°
5. 51°
6. 41°
7. 17°
8. 52°
9. 65°
10. 84°
11. 67°
12. 2°
13. 17°
14. 62°
15. 34°

Section 4: The Unit Circle: Signs of Trigonometric Values

1. A
2. B
3. B
4. D
5. C
6. A
7. B
8. D
9. C
10. A
11. B
12. B

13. C
14. D
15. A

Section 5: Graphing and Inverse Trigonometric Functions: Graphing Trigonometric Functions

1. Sine
2. Cosine
3. Tangent
4. Tangent
5. Cosine
6. Sine
7. Sine
8. Tangent
9. Cosine
10. Tangent
11. Sine
12. Cosine
13. Tangent
14. Cosine
15. Sine

Section 5: Graphing and Inverse Trigonometric Functions: Properties of Trigonometric Functions

1. 2π
2. ½ π
3. All real numbers
4. -2π to 2π
5. All real numbers
6. 4π
7. 2π
8. -2π to 2π
9. 4π
10. -1/5π to 1/5π
11. All real numbers
12. 2π

13. -2 to 2
14. All real numbers
15. 2

Section 5: Graphing and Inverse Trigonometric Functions: Inverse Trigonometric Functions

1. Cotangent
2. Secant
3. Cosecant
4. Sine
5. Cotangent
6. Secant
7. Cosine
8. Cosecant
9. Cotangent
10. Cotangent
11. Tangent
12. Secant

Section 6: Trigonometric Identities: Proving Identities

Each problem should show each step of the proof with the calculated value being very close to 1.

Section 7: Trigonometric Equations: Trigonometric Equations

1. x=0
2. x=45
3. x=90
4. x=270
5. x=180
6. x=0
7. x=90
8. x=45
9. x= 30
10. x=60
11. x=180
12. x=360

13. x= 30
14. x=60
15. x=270

Section 8: Oblique Triangle Laws: The Law of Sines
1. 846.42 units
2. 35.95 units
3. 66.85 units
4. 54.05 units
5. 17.36 units
6. 2.4 units
7. 468.23 units
8. 2.2 units
9. 214.65 units
10. 17.36 units
11. 144.65 units
12. 141.51 units
13.21.33 units
14. 44.64 units
15. 22.73 units

Section 8: Oblique Triangle Laws: The Law of Cosines
1. 4.31 units
2. 4.79 units
3. 30.56 units
4. 16 units
5. 162.73 units
6. 90°
7. 9.75 units
8. 7.74 units
9. 12.36 units
10. 10 units
11. 17.67 units
12. 58.87 units
13. 1.55 units
14. 119°

15. 14.24 units

Section 9: Complex Numbers and Polar Coordinates: Plotting Imaginary Numbers

1. To the right of zero ten, above the number line nine
2. To the right of zero four, below the number line seven
3. To the left of zero fifteen, above the number line two
4. To the right of zero six, above the number line five
5. To the left of zero nine, below the number line ten
6. To the right of zero twenty-eight, above the number line two
7. To the left of zero one, below the number line ten
8. To the left of zero nineteen, above the number line three
9. To the right of zero thirteen, below the number line one
10. To the right of zero fourteen, below the number line thirteen
11. To the right of zero twenty-one, above the number line fifteen
12. To the left of zero two, below the number line six
13. To the left of zero eleven, above the number line one
14. To the right of zero five, above the number line eleven
15. To the right of zero seven, below the number line eight

Section 9: Complex Numbers and Polar Coordinates: Combining Complex Numbers

1. 18+8i
2. 19+5i
3. 2+8i
4. 22-22i
5. -4-15i
6. 12-6i
7. -4+9i
8. 13+14i
9. 8i
10. 15+11i
11. 44+9i
12. -10-21i
13. -8+20i
14. 41+6i
15. -18+10i

Section 9: Complex Numbers and Polar Coordinates: Multiplying Complex Numbers

1. -85i+105
2. 72+30i
3. 100-75i
4. 14-135i
5. 37+37i
6. 23+24i
7. 104-216i
8. 52+38i
9. 4-124i
10. 175+20i
11. 23-11i
12. -22+54i
13. -38-41i
14. 46+22i
15. 3-29i

Section 9: Complex Numbers and Polar Coordinates: Complex Numbers in Polar Form

1. 7.21(cos (146.3)+ i sin (146.3)
2. 7.07(cos (45)+ i sin (45)
3. 7.28(cos (-105.9)+ i sin (-105.9)
4. 18(cos (40.6)+ i sin (40.6)
5. 35.7(cos (-153.4)+ i sin (-153.4)
6. 5(cos (-143.1)+ i sin (-143.1)
7. 13(cos (-22.6)+ i sin (-22.6)
8. 9.4 (cos (18.4)+ i sin (18.4)
9. 5(cos (90)+ i sin (90)
10. 3.6(cos (-33.7)+ i sin (-33.7)
11. 15.26(cos (58.4)+ i sin (58.4)
12. 15.6(cos (-39.8)+ i sin (-39.8)
13. 10.4(cos (-16.7)+ i sin (-16.7)
14. 27(cos (10.5)+ i sin (10.5)
15. 21.84(cos (-15.9)+ i sin (-15.9)

Made in the USA
Coppell, TX
03 August 2026

86213916R00085